Problems of Bioethics

Theologisch-Philosophische Beiträge zu Gegenwartsfragen

Herausgegeben von Susanne Dungs,
Uwe Gerber, Lukas Ohly und Andreas Wagner

Band 12

Frankfurt am Main · Berlin · Bern · Bruxelles · New York · Oxford · Wien

Lukas Ohly

Problems of Bioethics

Bibliographic Information published by the Deutsche Nationalbibliothek
The Deutsche Nationalbibliothek lists this publication in the Deutsche Nationalbibliografie; detailed bibliographic data is available in the internet at http://dnb.d-nb.de.

Cover Design:
© Olaf Gloeckler, Atelier Platen, Friedberg

Cover Illustration:
Sonography Scan.
Printed with kind permission
by Dr. med. Ulrich Schlomann,
Ortenberg. 2012.

ISSN 2194-1548
ISBN 978-3-631-63777-7
© Peter Lang GmbH
Internationaler Verlag der Wissenschaften
Frankfurt am Main 2012
All rights reserved.

All parts of this publication are protected by copyright. Any utilisation outside the strict limits of the copyright law, without the permission of the publisher, is forbidden and liable to prosecution. This applies in particular to reproductions, translations, microfilming, and storage and processing in electronic retrieval systems.

www.peterlang.de

*To the
St. John of Damascus Institute of Theology
in Balamand (Lebanon)*

Preface

This book is based on lectures I was holding in the St. John of Damascus Institute of Theology at spring 2010. Because of the strong ecumenical relationship between the Greek Orthodox Church (Syria and Lebanon) and the Evangelical Church of Kurhessen-Waldeck (Germany) I was invited to lecture about bioethical issues. I was grateful for the interest of audience and the hospitality of the members of the institute during the trimester.

The lectures develop an ecumenical perspective on several issues of bioethics. Therefore I use arguments which transcend the perspective of a special Christian confession. The main focus is to reveal religious implications of hopes and scenarios in biomedicine and biotechnological research. Often these scenarios are similar to theological presuppositions in form or even content. It seems to me that religious implications are an implicit part of a new science, so that there need to be no contradictions between faith and science in the technological age, at least in principle. What I want to show is that no actual bio-ethical position can avoid religious or metaphysical implications even if it might seem possible.

It seems to me that religious philosophy can build a common platform which might support theologians of all confessions arguing with one voice about bioethical innovations. Thus, these lectures of a protestant theologian in an orthodox faculty were a challenge to build such a platform.

For correcting my papers I thank Daniel Fitzpatrick whose linguistic skill supports the publishing as strong as his philosophical empathy. A special thank to the listeners of my lecture – students, scientists, alumni and priests – whose questions and objections helped me to develop my arguments more precisely. One of them – my friend Hector Abouid Bandy – was correcting passages of my manuscript. I am very glad for his advice and powerful support. Bishop Prof. Dr. Martin Hein of the Evangelical Church of Kurhessen-Waldeck supported me in organising my visit. I am very thankful for that.

September 6th 2012   L. O.

Contents

Preface 7
1   The claim of bioethics 11
    1.1   Three suggestions 12
    1.2   Appendix: The principles of ethics 16
2   The moral state of the embryo 21
    2.1   SCIP-Arguments 24
        2.1.1   The Species-argument 26
        2.1.2   The Continuum-argument 26
        2.1.3   The Identity-argument 27
        2.1.4   The Potentiality-Argument 28
    2.2   An argument against the human state of the embryo 30
3   The power of feelings in bioethics 37
    3.1   Feeling, Wholeness and Nature 37
    3.2   How does biotechnology influence our feelings 43
    3.3   Why do scientists have no scruples? 45
4   Human reproductive cloning and germ line therapy 49
    4.1   Reproductive Cloning 49
    4.2   Germ line therapy 53
    4.3   An anthropological framework to evaluate cloning and germ line therapy 56
    4.4   How genetics disturbs human freedom 62
    4.5   Result 67
5   Brain and Mind 69
6   Could computers feel like humans? (Qualia) 77
    6.1   First Case 78
    6.2   Second Case 80

- 6.3 Criticism 81
- 7 Patients with serious brain damage 87
- 8 Do Humans have a free will? 95
  - 8.1 What is free will? 95
  - 8.2 The Libet-Experiment 97
  - 8.3 Freedom means self-determination 100
  - 8.4 Theological appendix 104
- 9 The problem of mercy killing 107
- 10 Eternal Dignity 117
  - 10.1 A metaphysical presupposition of Dignity 117
  - 10.2 A second metaphysical presupposition of dignity 120
  - 10.3 Non-metaphysical approach to end of life-decisions 122
  - 10.4 How metaphysical Dignity works in Ethics 129
- 11 Ethical and theological kinds of guilt 139
- 12 The Metaphysical Concept of Presumed Will 145
- 13 At the End of the Lectures 151

1 The claim of bioethics

In December 2008, I was glad to have been invited to visit Balamand the first time in my life. There was a public conference with the title "Bioethics. The Need of Paradigm Shift?" The question was: Do we need a new kind of ethical thinking? Do we need new strategies of decision-making? Why did this question arise? It was because of the fundamental technological revolution in biomedicine and biotechnology.

There are a lot of fantasies which arise from new technological possibilities. There is the fantasy of immortality, the fantasy of creating life, the fantasy of reproducing life by robots. Most of these fantasies are science fiction. It is far from clear if they will come to reality. If they do, it will have consequences for the meaning of life, death and immortality. It will also have consequences for Christian thinking of resurrection and eternal life.

What is not a fiction, is the present claim of bioethics. Bioethics claims to develop new kinds of ethical thinking. Although it is far from clear if many fantasies of biotechnology will be realized at all, bioethics suggests to define the moral frameworks of a new world, which might never come. The meanings of basic terms of our self-understanding change, in spite of the uncertainty whether the social arrangement could endure such a shift.

But the main problem I see in that development is the following: Bioethics suggests that ethics is a matter which could be simply changed like technical methods. Bioethicists understand ethics as technical issues. The term "paradigm shift" derives from the scientist Thomas Kuhn (1922–1996). Kuhn has shown that sciences develop their works within a paradigm. And they defend the basic paradigm, even if contradictions arise between the paradigm and concrete results of scientific research. When this happens, the researcher may have made a fault, but the paradigm remains true. Only if the difficulties between paradigm and research arise, which could never be integrated in the paradigm, is there a force to shift the paradigm. Thus a paradigm shift is not a desirable situation even in science. But sometimes scientific or technological demands urge such a paradigm shift in order to improve the scientific results.

1.1 Three suggestions

It seems to me that bioethics suggests firstly that a paradigm shift in ethics is desirable, whereas in science a paradigm shift is avoided as long as possible. This is the first difference between bioethics and science. Secondly ethics becomes a technical issue by the term "paradigm shift". I am skeptical whether ethics is able to turn to another paradigm simply because therefore we would need moral reasons to turn. And these moral reasons would have to be the traditional ones. And this would exclude a paradigm shift. In my opinion ethics could revise judgments, could develop judgments, could describe moral problems more precisely. But ethics could only do that within the scope of ethical phenomena. And this scope of ethical phenomena includes a necessary kind of making judgments. There is only one paradigm in ethics or there will be no ethics at all. There is no possibility of paradigm shift. And this is the second difference between ethics and technical issues.

This leads to a third suggestion of bioethics. This is the suggestion that ethics could only be handled in different areas. There are no further common ethics. There is no basic framework for ethical judgments in general, which need only be translated in different areas like medicine, politics, science and social life. Instead of this bioethics suggests that it deals only within this one area and from this area. And it suggests that this area is separated from other areas of life. Bioethics would not like to be ethics because it refers to a common frame of ethical principles which is valid in all areas of life. But bioethics would only be ethics because it merely refers to the special problem of bio-medicine and biotechnology. Bioethics is the claim of a kind of "area ethics" which loses its integral appearance. The type of arguing is the image of an "intrinsic logic". Area ethics suggests that there are no common rules for moral decision making. But every decision will be made by the intrinsic logic of a special ethical field, for instance bioethics.

The German theologian Traugott Jähnichen justifies this development of differentiating ethics in plural areas. According to Jähnichen there is no common normative link between these areas. Rather all areas are autonomous in defining the morality in their issues. And they are autonomous in solving their problems by their own constructions of ethics. This does not mean that ethical judgments become impossible – as "ethical" judgments. Therefore Jähnichen rejects opposite strategies to react to differentiated social areas: On the one hand he rejects the transformation of ethics into private matters, whereas the public area remains moral-free. On the other hand Jähnichen rejects the claim that a differentiated area runs only by its own rules and leaves no room for ethical judgments above the area. According to Jähnichen, this would be a kind of determinism which does not describe the area well. Economy, politics, science, technology and medicine have no determinative definition how each of them functions. All social areas function differently than logically or causally determinative[1]. They are not closed systems. Thus they leave room for ethical negotiations which might be strange for them initially.

Thus Jähnichen proposes a third way of ethical decision making within a special area, instead of ethical judging from an outer perspective on the one hand or destroying ethics on the other. Jähnichen's interest is to incorporate ethics in a special area in order to control the autonomous run within the area. Area ethics has to develop an analytic look onto the phenomena of the special area and to describe the intrinsic logic of its functional rules. Intrinsic rules could be evaluated by ethical analysis, simply because they are not in opposition to ethical analysis *in itself*. Intrinsic rules could be ethically acceptable or even desirable. It could be ethically useful to differentiate the social order into different areas

---

[1] T. Jähnichen: Wirtschaftsethik. Konstellationen – Verantwortungsebenen – Handlungsfelder; Stuttgart 2008, 50f.

like politics, economy, science, technology and so on. It could be ethically right to live in such a social order of functional differences in several areas. And it could be ethically right to improve the autonomous ruling in every area.

So, according to Jähnichen, bioethics is a justified kind of area ethics. Simply because the technological revolution of the last decades runs according to its own rules and according to its own intrinsic logic, new ethical problems arise which could only be solved in an incorporation of ethics in the intrinsic logic of that area of biotechnology. We need bioethics because biotechnology functions according to its own rules.

I would object: How could we identify what is ethically acceptable or even desirable if we had no extrinsic measures of evaluating? How could we identify whether something, which is technologically interesting or aesthetically attractive, is also ethically right or good? How could technical issues give us an impression what is ethically right or good? It seems to me that Jähnichen's proposal is circular: It already includes what it wants to prove afterward. Jähnichen wants to show that ethics is already implanted in social areas; but therefore he needs the presupposition that everyone within that area already knows what ethics is. However, what ethics is could only be defined by the intrinsic logic of a special area for itself. The result is that there is no intrinsic criterion why special types of controlling an intrinsic process within an area are ethical and not, for instance, technological or economical or political. By this way, we lose the criterion how to distinguish ethics from other ways of decision-making or describing processes. Rather the intrinsic logic of technology describes what is ethical. And this is ethically unsatisfactory because it eliminates the autonomy of ethical thinking.

Another young German theologian is Philipp Stoellger. Stoellger also objects the claim of area ethics in the way Jähnichen does. But Stoellger argues from a strict theological perspective. In an article from 2009, Stoellger calls the process of differentiating theological thinking in several areas the "end of theology"[2]. Theology sacrifices its own kind of thinking on the altar of technology, science, political pressure and economic demands. Thus theology loses its own resources of understanding what happens with life by the biotechnological revolution. This is not a technological question but a question of understanding. And according to Stoellger, understanding is the basic task of theology.

Understanding is the move from myself to the other. Understanding is not only the other. Since I understand someone I do not become the other. I remain myself. And I could only understand by remaining myself. But in understanding I move from myself to the other. I am obligated by another person. But I am obligated as myself[3]. And this means that I transform the other for my perspective. Stoellger mentions that: "Whoever understands, understands for himself and so understands generally something different (than what is said by the other)"[4]

In this way ethics – at least theological ethics – is necessarily in face of areas but never within. Ethics needs to be independent from technological development. It is not ethical thought since one thinks biotechnological innovation would change our values. It is not ethical thought, since one thinks biotechnology develops its own new ethics which has nothing to do with moral traditions. It is a loss of ethics, since one thinks ethics of life could only be bioethics.

To sum up: On the one hand it is right that the biotechnological revolution produces new problems of understanding questions like "What is life?", "When does life begin?", "When does it end?", "Could life be saved by technological methods although persons are already dead according to classic understandings of life?" and so on. New problems arise. We have to contemplate new social and technological contexts in which ethical problems arise. Insofar as we have ponder "bio-ethical problems". I accept the term bioethics as long as it refers to such

---

2   Ph. Stoellger: Missverständnisse und die Grenzen des Verstehens; ZTHK 106/2009, 223–263, 225.
3   Ibid, 250f.
4   Ibid, 262.

problems and as long as the term does not support the prejudice that ethics needs a paradigm shift.

On the other hand, I see a dangerous suggestion within the bio-ethical claim that everything becomes different. It might be that the biotechnological revolution sets new facts in social life. It might also be that this revolution suggests new values or new kinds of thinking. But this does not mean that the traditional kinds of evaluating *should* change. We have to argue *by ethical rules* whether new kinds of thinking are justified. Therefore we need ethical arguments, ethical principles, ethical strategies for evaluating. Such is to say: We need features to identify what actually is ethical. And these features are the traditional ones, the ones which make the essence of ethics. And the essence could not change.

1.2 Appendix: The principles of ethics

But what is the essence of ethics? Do we have an independent perspective, independent from our prejudice, independent from our personality or our history? If "No", then we do not have any key to persuade others with a different prejudice, personality or history. Then the only way to persuade someone what is ethically right would be to force him to convert. But how could we *know* that our ethical perspective is the right one? Could not we only believe that? And since we only believe, then our ethics is unsatisfactory: Our ethics is right because we believe in it. And what is right depends on what we believe to be right.

Many Christians justify their ethical claims by authority: This (or that) is right because a divine authority has called it right – for instance God himself or a revelation or an authority of the church. Behind that, there is the presupposition that the creator has not only created the world but also essences, especially the essences of ethical judging. If God had called something different as right or good, because of his free will, then this something different would *be* right or good. The Divine being would have had the freedom to define something as right or good, which actually is bad or evil or wrong. So the difference good/evil, right/wrong becomes a difference of divine arbitrariness, since one maintains that ethical claims are sufficiently justified by authority.

Here we need an important difference between morals and ethics. Morals are the tradition or the traditional agreement or the common sense about what is right/wrong and good/evil. It depends on lifestyle and social commitments which have been built during the history. We know that morals change fluently. There are plural influences in our society why we change our judgments about moral matters. And we have an increasing pluralistic situation in which everyone develops own attitudes or individual understandings of what seems to be common sense.

In opposition to that, ethics is a kind of *reasoning*. Ethics checks the moral claims or the alternative recommendations of what should be. And ethics prefers such claims which are justified at best. But what is a good reason? What is a good justification? It seems to me that a good reason has to do with the essence of ethics, that is reasoning. This sounds circular but it is not. The best reasons or the strongest reasons are such which take care of the situation which is given by reasoning. Whoever justifies claims, respects his neighbour to whom he speaks. The neighbour has the right to get justification. That means the neighbour is respected as equal as the person who justifies. This is because there is a law of reciprocity: When I justify claims, my neighbour has also the right to justify what his claim would be. And this constitutes the condition of equality in ethical discourse.

The neighbour has the right to get a justification he could accept, even if he has different beliefs or attitudes. His equality is independent from attitudes and beliefs. And this has consequences about the best reasons in ethics: The strongest ethical reasons refer to the situation of reasoning. Since every person has the right to get reasons for moral claims it is clear that the strongest reasons are such which are most understandable. Now, what is most understandable, does not only depend on the speaker. Rather it usually depends on the listener: What can the listener understand? Would she have enough resources to understand? Does she understand the language, the words, and the logical structure of an argument? But since the listener has the right for getting reasons independently from her own attitudes and beliefs, the consequence will be the following: The strongest arguments are such which need no special attitudes or beliefs in order to become well understood.

For instance, it seems not to be a very strong argument to justify a special norm by a divine authority. "You shall not kill" is not right because God has told us not to kill. Everyone who does not believe in God could not understand why this should be a meaningful justification of the commandment. And if the only reason not to kill refers to God's giving this commandment, everyone could reject that claim who does not respect a divine power.

Theological ethics does not demand that everyone must convert to Christianity in order to understand or even accept the claims of Christian ethics. Theological ethics allows everyone to remain what he or she is and which attitudes and beliefs he or she has. The purpose of ethical discussion is not converting the other but of persuading the other who might remain the other with his or her different attitudes. So I would like to highlight the fact that theological ethics is also ethics, that means it has the same properties, the same features like ethics in general. Also theological ethics is reasoning. And it has its strongest arguments, since its reasoning is independent from attitudes and beliefs. The weaker the presuppositions are the stronger the arguments become. This is also valid for theological ethics. Theological ethics fulfills the same properties as ethics in general. Otherwise it could not be identified as ethics.

This does not mean that theological ethics should avoid its special Christian perspective. But this is the interesting point of theological ethics: How to translate Christian values in an argument Non-Christians could share? How to translate Christian moral standpoints to arguments others could accept without converting their religious beliefs?

Such at least theological ethics but perhaps ethics in general is a kind of "cross over". Theological ethics is a strategy, a creative entry in communication in order to hold Christian claims in a non-Christian context. Thus theological ethics is a way – perhaps the only way in theology – to communicate Christian truth claims as acceptable for Non-Christians. Not converting but persuading is the purpose of ethics.

When I speak about the principles of ethics, I mean firstly this context of reasoning. Even in times of bioethics, we need to give reasons for our claims. This will not change. With the term "principles of ethics" I do not mean basic values such as dignity or justice. Nor do I mean ends of ethics, higher or highest values such as love or eternal peace or enhancement of human conditions. It might be that all these values need transformation during historical events or epochs. Thus bioethics might enforce such a historical change of ethical terms. But what I mean, is that the context of arguing never changes ethics. Ethics is reasoning or there is no ethics. Ethics presupposes equality of everyone who could take an ethical position. This will never change or there is no ethics. Ethics presupposes that we are responsible for deciding what to do. There is no intrinsic logic of an area which makes everything determinative what happens and which destroys our responsibility. Responsibility is one necessary condition of ethics and thus a principle of ethics. And this will never change or there is no ethics.

In my opinion these are the three principles of ethics: Social embedding of reasoning; the equality of everyone who is concerned to ethical decision-making; and responsibility which includes freedom to act. Suppose a situation in which humans lose their freedom, their responsibility to decide what they should do. Such a situation would entirely change our ethical thinking, or better: it would destroy ethics. Thus there is no ethical reason to enter such a situation. It might be that under some biotechnological circumstances, humans lose their conventional capacity to decide ethically, for instance if their human dispositions are programmed by a brain manipulation which deals with humans like with robots. Some day, it might be technologically possible to achieve such kinds of situations. But such kinds of situations could never be ethically justified. They could only be a horrible accident of social life. But they could never be justified. Otherwise it would be ethically justified to leave ethical reasoning. And this would be a contradiction of ethical thinking in its entirety.

When I discuss problems of bioethics in this book, I will show which problems arise by biotechnological innovations. But I reject the claim that the principles of ethics have disappeared under such circumstances. I will show when such a claim is held. But it is not necessary anymore to give reason why this claim is false, because the contradiction is already obvious.

2 The moral state of the embryo

Now I turn to the first example of bioethics. Abortion and contraception are one of the oldest bio-ethical subjects and has been discussed since decades, almost one century or even longer. It is a bio-ethical subject because it changes the meaning of life as seen technically. If life begins with birth, it is logically impossible to kill a human before birth. But technically, it has become possible to kill a fetus. During the last century, it has become possible to do that without serious risks for the pregnant woman. So the question about the moral state of a living fetus could become relevant.

Since some decades, contraception has become an improved practical way for having sexual intercourse without becoming pregnant. This technical innovation has changed our understanding about sex. And it also refers to the understanding of life. Is contraception a kind of killing or at least destroying human life? In all these cases new ethical problems have arisen by new technical possibilities.

Since the eighties of the last century, there has been a further innovation, that is conception without pregnancy. This innovation is called In-Vitro-Fertilisation, in short IVF. It is possible to extract an egg from a female uterus and a male sperm and to unite them in vitro, beyond sexual intercourse and outside the body of a female. Such IVF is conception without pregnancy. Until now, it is not possible to develop a human embryo outside the womb. An embryo could survive only some days outside the womb. But despite this, the innovation of IVF has serious consequences for our understanding of human life before birth.

I would like to list some practical consequences:

1) Through IVF, it becomes possible that the embryo has two biological mothers: The donor of the egg and the woman for pregnancy. It is possible to implant the embryo in the womb of a different woman than the one who has given the egg.

2) Through IVF, it is possible to reject an implantation. This is the case of conception without pregnancy. The egg could have been fertilized. But after

conception the mother changes her wish to become pregnant. In this case, the embryo must be destroyed or frozen. The frozen embryo could survive for years, at least in principle. Then it could be adopted by another woman or the original donor could change her wishes again and could agree to implantation.

3) Outside the womb, it is easier to test the embryo whether there are genetic defects or disease. This is called "preimplantation genetic diagnosis" (PGD). In some cases the defect might be corrected. But in most of these cases the implantation would be rejected. So IVF is a comfortable technology for prenatal medicine. But on the other hand it is also the beginning of eugenics and a threat of discrimination against ill or handicapped infants.

4) IVF makes research with embryos possible. Since embryos are outside the womb one could explore their genetic structure in order to find genetic defects or to develop new therapeutic means against genetic-based sickness. Cloning also becomes possible. One could differentiate the embryonic cells in order to multiply them in several biological systems. The embryonic cells have a special property which get lost in further embryonic development. This is the property so-called "totipotence". Totipotence means: The cell could develop to every body cell: skin or brain cell or blood or of other organs. Totipotence is an important property for medical research with human cells. So scientists need more totipotent cell-material. Unfortunately by extracting the totipotent cell in the earliest stages of the embryo's life the embryo dies. Thus research of embryonic cells destroys embryos. Thus embryo research reveals an ethical ambivalence: On the one hand the embryo in vitro offers the possibility to establish new totipotent cells by cloning. On the other hand research of totipotent cells destroys embryos.

These cases provoke the question how to evaluate such a technology. Usually, most of the ethicists reduce the question of evaluating these cases to the question of the moral state of the embryo. For me it seems to be a reduction because IVF could also be ethically problematic even if the embryo has no moral state. The question about the moral state is only one question. But the moral state is neither necessary nor sufficient for evaluating this new method of IVF. It is not necessary because IVF could remain problematic even if the embryo has no moral state. And it is not sufficient because there could be good reasons for the practice of IVF even if the moral state is a very high value, e.g. the level of human dignity.

A further reduction is the interpretation of the moral states of the embryo as its ontological state. "Ontological" means the doctrine of the being. Most ethicists discuss the question about the moral state of the embryo by clarifying if an embryo is a human being. Since an embryo is ontologically classified as an example of humankind, it deserves the same moral state as all humans. So the question about the ontological state of the embryo could solve the problem of the moral state of the embryo.

In my opinion, this is also a reduction. Like above the ontological state of the embryo is neither necessary nor sufficient for clarifying its moral state. Suppose we could prove that an embryo is a human being. Does that mean that an embryo has the same rights as other humans? It seems not because its existence is not entirely independent from the mother. And fundamental interests of the mother could conflict with interests of the embryo. In such cases, it seems morally responsible to place the interests of the mother over those of the embryo, at least because of logical reasons: If the fundamental life conditions of the mother are harmed, the embryo would also be harmed[5], but not vice-versa. So the rights of the mother seem to be founded more deeply than the rights of an embryo. But if so, it seems to me that the ontological state of the embryo is not a sufficient condition for its moral state.

Moreover, it is not a necessary condition. Suppose we find out that embryos are not human beings. It does not follow then that they have no moral state at all. According to the bioethicist Giovanni Maio, the embryo could be valuable in

---

5 Unless one would extract the embryo from the womb in order to implant it to another woman. But in this case there is no duty of another woman to get the embryo implanted. The consequence is that the embryo has no right to become implanted anywhere.

other references, e.g. the embryo could be of instrumental value (without embryos there is no reproduction of human mankind); the embryo could play a special social role in society we should respect; the embryo could represent human borderlines we should respect in order to hold an existential balance to ourselves[6]. Thus the statement that the ontological state of embryos is different from the ontological state of humans does not lead to the conclusion that embryos have no moral state at all.

So in my opinion, the main line of the discussion about the ethical estimation of IVF and prenatal research leads to a double reduction. The reason for this might be the bio-ethical suggestion that everything becomes different because of the biotechnological revolution. Bioethics suggests that strict ontological lines become fluent. Therefore the classic ethics seems to collapse. This is an analytical mistake. We could say that the ontological state of embryos needs to have nothing to do with its moral state. And the moral state of the embryo needs to have nothing to do with the ethical estimation of the IVF-practice.

Despite this, I would like to draw out the present discussion about the ontological state of the embryo. This discussion shows how bio-ethical arguments function, namely as a combination of science fiction and logical analysis.

2.1 SCIP-Arguments

The artificial term "SCIP" refers to four typical arguments why the embryo has the same ontological state as a human. Each letter represents one type of arguments:

1) **S**pecies-argument
2) **C**ontinuum-argument

---

6 G. Maio: Ethik der Forschung an verwaisten Embryonen – Erläuterungen zum Respektmodell; Bioethica Forum 37/Juli 2002, 22–30, 25.

3) **I**dentity-argument
4) **P**otentiality-argument

The Species-argument is the following: Everything which is a member of the humankind is a human being. Thus all embryos are humans.

The Continuum-argument is the following: The human development has no breaks or gaps. Everything to what the human develops is entailed in him forever. Thus: When humans have dignity, then also all embryos have dignity because they develop to humans continuously.

The Identity-argument is the following: An embryo is the identical being as a human. Thus it has the same dignity as a human.

And finally the Potentiality-argument: An embryo is potentially a human. Thus although an embryo has not still the same ontological state like humans it deserves the same rights like humans because it has the full potentiality to become a human.

In the year 2003 the two German philosophers Gregor Damschen and Dieter Schönecker have published a book in which they destroy the plausibility of all the four arguments of SCIP. Their only check was the logical analysis of the SCIP-arguments. Despite their result that all SCIP-arguments lapse, they accept a combination of SCIP as a rational argument for the moral state of an embryo. Although SCIP could not really prove what an embryo is, it makes it questionable whether an embryo is *not* a human. And in this situation, Damschen and Schönecker argue for protecting embryos because perhaps they could be humans. And that is the headline of their publishing: "In dubio pro embryone" – In Doubt For the Embryo.

Now let us see why the SCIP-arguments are not valid according to Damschen and Schönecker. Let us begin with

2.1.1 The Species-argument

Reassure the Species-argument: Everything which is a member of humankind is a human being. Therefore, all embryos are humans.
The objection against this argument consists in the following: Since it is clear why humans have dignity, the species-argument may function. But perhaps the moral state of humans is based on special properties not possessed by the embryo. Then the species-argument is not valid.

This leads to the main problem of the species-argument: The membership in the humankind seems to be more important than the moral properties of humankind. But then the ascription of dignity to the human mankind seems to be arbitrary.

2.1.2 The Continuum-argument

Regarding the Continuum-argument: The human development has no breaks or gaps. Everything to what the human develops is entailed in him ever. Thus: When humans have dignity, then also all embryos have dignity because they develop to humans continuously.
According to Damschen and Schönecker this is the weakest argument of the SCIP. They object that it is not an argument at all but only a view. Their question is: What exactly would actually a continuum be? One proposal could be: A continuum is a spatial-temporal unity. The Continuum-argument seems to suggest: Only because the embryo is a spatial-temporal unity and it develops to a human in that time, it is already a human.

Also in this argument, the ontological suggestion ignores possible significant moral steps in development. Also if the embryo is a continuum, it could happen that during its development significant changes occur, which are morally relevant. Although the continuum may remain a continuum, it is possible that the ontological state of the continuum changes. The Continuum-argument would like to suggest that an embryo is not only a continuum but a human continuum. However the continuum-argument will prove that simply by showing that the embryo is a continuum. And this is not sufficient.

2.1.3 The Identity-argument

Regarding the Identity-argument: An embryo is the identical being as a human. Thus it has the same dignity as a human.

Also here the connection of moral and ontological implications is puzzling. It might be that an embryo is identical to a human. But does it follow then that identity is the *criterion* for the moral state of the embryo? It seems not. The embryo could not have the same dignity as a human because it is identical with a human. But an embryo could only have the same dignity as a human because it shares the same moral properties as a human.

Let me give an example to illustrate the difference. I am a theologian because I have passed an exam for theology. This is the sufficient property which makes me a theologian. That is, I am not a theologian *because* I am identical with a theologian. The latter explanation would fail to say what a theologian actually is. The ontological question has to be what makes me a theologian. And this is not answered by the statement that I am identical with a theologian.

That is the reason why it is not sufficient to call an embryo a human *because* it is identical with a human. The Identity-argument ignores the problem: what is the significant feature of being a human? So the Identity-argument fails.

2.1.4 The Potentiality-Argument

Regarding the Potentiality-argument: An embryo is potentially a human. Thus although an embryo has not still the same ontological state like humans it deserves the same rights as humans have because it has the full potentiality to become a human.

I assume you will feel the same as I do: The Potentiality-argument identifies entities although it concedes that they are not the same. The Potentiality-argument concedes that an embryo is not a human but it could only become a human. And this is a very weak argument for ascribing human rights for something which is actually different but only potentially becoming the same.

What exactly does potentiality mean? The term potentiality could have a stronger and a weaker sense. Has it a weaker sense, then almost everything could become a human. Atomic structures of other beings could become incorporated in a human body. Technical entities like a prosthesis or a pacemaker are potentially a part of a human. And this is the case for the matter of these entities. But to say that everything deserves the same rights like humans because it could happen that it becomes something human, is ridiculous.

In a stronger sense, potentiality must exclude weaker forms of potentiality. For instance, if an embryo is potentially a human then also the unfertilized egg is potentially a human and also a sperm. But this would lead to absurd consequences: It harms the egg not to fertilize. It harms the sperm to masturbate. Masturbation would be a kind of mass-murder, but also monastic life. When we want to avoid such absurd consequences we must define a stronger criterion what potentiality actually means.

Damschen and Schönecker propose the following description of potentiality in a stronger sense: *potentiality is the actual ability to establish all moral-relevant skills later on*[7]. And now Damschen and Schönecker turn their analysis

---

7   G. Damschen/D. Schönecker: In dubio pro embryone. Neue Argumente zum moralischen Status menschlicher Embryonen; in: G. Damschen/D. Schönecker (Ed.): Der mo-

and develop their argument "in dubio pro embryone". They accept that potentiality in their interpretation is valid. The embryo is indeed actually able to establish all moral-relevant skills later on. But this does not make an embryo a human. Thus Damschen and Schönecker combine their potentiality-argument with the identity-argument. Why? Because the unfertilized egg and also the sperm have the same kind of potentiality. Both are also actually able to establish all moral-relevant skills later on. But they are not identical with the embryo. However they need to be transformed to an embryo in order to establish their moral-relevant skills.

Thus an unfertilized egg or a sperm are not quantifiably identical with an embryo. And so they could not be quantifiably identical with a human being. This is different to an embryo. An embryo is indeed quantifiably identical with the human being it potentially will be. Thus it is sufficient to identify an embryo quantifiably with the human it potentially will become. However, the identity-argument fails to ascribe human rights for an embryo. But according to the both German authors, the identity-argument is valid in combination with the potentiality-argument. Now it seems strong enough to bolster skepticism against the position that embryos might not be human.

The skeptical strategy in ethics Damschen and Schönecker use, is called "Tutiorism". Tutiorism means: We hold a stricter protection than we actually give reason for, because we could not exclude the possibility that embryos are humans. In a situation of uncertainty, we accept the stricter norms, we respect protection laws with a higher level. The reason is: If it is possible that embryos are humans then we must protect them. But if they are not humans there is no harm against them to protect them like humans.

---

ralische Status menschlicher Embryonen. Pro und contra Spezies-, Kontinuums-, Identitäts- und Potentialitätsargument; Berlin, New York 2003, 187–267, 227.

2.2 An argument against the human state of the embryo

In 2005 and 2006, I have published an argument against the human state of the embryo[8]. I have not argued in this way in order to weaken protection of the embryo. My argument was not held for unlimited liberty of research. But I want to show that the crucial point of the ethical problem is something else: It is not the ontological state of the *embryo* but *our* moral feelings and intuition which we actually should discuss. So in the next two sections I would like to show both: In this section I draw out my argument why embryos are not individual humans. And afterward I want to show what the problem of destroying embryos has to do with *us*, with our *morality*.

The core idea of my argument consists of the fact that human dignity and human rights are individual rights. But embryos are not still individual humans. They might be individual embryos; even they might be human individuals – like organs or parts of the body. They are human individuals, but they are not individual humans, that means a complete human, who is, what he is, as an individual human. Therefore, we must leave the bio-ethical discussion shortly for introducing a philosophical remark.

In the seventies, the American philosopher Saul A. Kripke proposes a criterion for identifying something as the same. This criterion is the link to the first act of referring to something or someone. For instance, I identify Aristotle by referring first to the act of referring to Aristotle. With Aristotle I mean the person someone has meant when she has referred to him for the first time. Kripke's criterion might seem to be unnecessarily complicated, but it is helpful in order to avoid serious philosophical problems. And it becomes apparent when one has understood the core idea of Kripke.

---

8 L. Ohly: Das Erschrecken vor sich selbst. Therapeutisches Klonen als Spiegel für das menschliche Selbstbewusstseins; in: P. Dabrock, J. Ried (Ed.): Therapeutisches Klonen als Herausforderung für die Statusbestimmung des menschlichen Embryos; Paderborn 2005, 151–167; L. Ohly: Konkrete Embryonen und konkrete Menschen – Kripkes Tipps zur Vermeidung einer Irritation; Zeitschrift für evangelische Ethik 50, 2006, 277–290.

One problem which could be avoided is the imagination of something as the class of its own properties. Suppose we call Aristotle the last important ancient philosopher. Here we link up a essential property with a person and define that person by a special property: the last important ancient philosopher. Now suppose that historical research finds out that after the death of Aristotle there was another important philosopher in the ancient world, someone whose scriptures were disappeared for a long time until they were rediscovered again in present times. According to the classical theory of identification by properties we would have conceded: This is Aristotle – and not the one we have had in mind until now. Not the writer of Metaphysics, not the most famous scholar of Plato and so on.

But then, could not Aristotle be a bundle of several properties? The last important ancient philosopher, the writer of Metaphysics and the most famous scholar of Plato? Suppose one of these properties is wrong. But the set of linked properties is still valid. This is the theory of the philosopher Peter Strawson: According Strawson, something is individually identified by referring to the bundle of properties which are linked up as disjunctions. For instance, Aristotle is the last important ancient philosopher *or* the writer of Metaphysics *or* the most scholar of Plato. Aristotle needs not to have all these properties but at least one or some of them. And he is identified by the link of these disjunctive properties.

Kripke rejects also this theory. He maintains that this link is arbitrary; it is too complicated and contradicts our ways of identification in daily life; and the bundle-theory is wrong in the end. The simple reason is the following: Given that any entity possesses the most properties of a bundle, and more properties than another entity possesses, it is not necessary to conclude that the first entity is the true referent[9]. Suppose there is another scholar of Plato, let us call him Smith. Although Smith has not written Metaphysics, he was the last important ancient philosopher historical research might have discovered. So the bundle of Aris-

---

9   S.A. Kripke: Naming and Necessity; Malden 1981, 52f.

totle is valid for Smith too. Could not we say then that Smith is Aristotle? Or must we even say that?

Kripke avoids such consequences which contradict daily life communication. Thus he develops a practical criterion how to refer to something. And this criterion is the link to the first practical situation in which someone has referred to that entity. Kripke's criterion leads to the consequence that *the actual conditions for the first reference-act of an object are still valid for all counterfactual situations of reference of that object.* This is a very important consequence. Aristotle remains Aristotle even if historical research would discover that someone else has written Metaphysics or even if another person fulfills the bundle of properties better than he. Even if Aristotle was in many ways different than we think of him or even if Aristotle could have been different than we think of him, he remains Aristotle. This is significant in Kripke's thinking: to establish different possible alternatives, which someone could have been without changing his identity. The reference to him remains true, even if many other situations could have happened. One calls this thinking "possible worlds"-scenario. And Kripke is concerned by the stable factors of different scenarios of possible worlds. Aristotle remains Aristotle also in different possible worlds, even in those in which he had not lived.

But the consequence of Kripke's reference-criterion has another effect: I mentioned that *the actual conditions for the first reference-act of an object are still valid for all counterfactual situations of reference of that object.* That means we would not know who we mean, if the original conditions of being Aristotle had changed. Kripke gives an example: For his possible world-scenario it is possible that a child of family Truman could have become the Queen of Great Britain. But it is entirely impossible to think that Trumans could have been the parents of Queen Elizabeth II. The latter is impossible simply because the conditions for being Elizabeth are her real parents. It is impossible to speculate about Elizabeth's parents different from her actual ones because her factual parents are necessary conditions for us to refer to Elizabeth. Her real parents are linked up to her significant reference[10]. This is simply because the rule is valid: *the actual conditions for the first reference-act of an object are still valid for all counterfactual situations of reference of that object.*

My last remark is important for my argument why embryos are not individual humans. The point is that we could not signify them like individual humans when we refer to them. Therefore, I establish a logical conclusion from several premises.

Premises:
1) No individual human can be born without a woman who bears him.
2) Every individual human can be born only by one woman who bears him.
3) By IVF the embryo could be implanted in different wombs.

precondition
p1) I call an individual human Tim who is born by Inge.

Conclusions:
C1) Tim could not have been born by someone else than Inge.

p2) I call an individual human Rüdiger who has the same embryonic origin and could have been born if the embryo had been implanted in Anke's womb instead of in Inge's womb.

C2) It is impossible that Tim is the same individual human as Rüdiger.
C3) It is necessary that the existence of Tim excludes the existence of Rüdiger, and vice-versa.

---

10 S.A. Kripke: Naming and Necessity; Ibid., 112f.

C4) The embryo who was implanted in Inge's womb, so that Tim was born, is not identical with Tim. The actual conditions of referring the embryo are different from the actual conditions to refer Tim.

C5) If Tim has a right to live, then Rüdiger has necessarily no right to live, and vice-versa.

Explanation:

On C1:

Even in a possible world in which Tim was not born, he could have had born only by Inge. The reference-link to Inge is necessary in order to refer to Tim.

On C2:

At least in some possible worlds the difference between Tim and Rüdiger are significant, namely such in which either Tim or Rüdiger are born. The original conditions of reference are different between Tim and Rüdiger, so that they could never be identical.

On C4:

Although Tim could speak of himself in past when "he" was an embryo he could only do that after referring to the first act of reference. And the first act of referring to Tim is his birth. However one could object that the first reference is an arbitrary act. Why could not we refer to Tim originally when we refer to an embryo? I concede that calling a name is arbitrary. And I concede that the act of naming is arbitrary. But every arbitrary reference-act has rigid consequences. The reference-act leaves no room for interpretation of what we mean. The rigid consequence is: Under alternative referential circumstances we would mean someone else than Tim.

Suppose we have already called the embryo Tim. Then we would mean someone else than the person we mean since we called him Tim after birth. The actual conditions of referring the embryo are different. For instance Tim cannot speak. He cannot breathe. It is impossible to refer to an embryo and to ignore the fact that an embryo could not speak or could not breathe. It is impossible that embryos speak or breathe. But then the human we have called Tim after his birth is someone else, because his actual conditions for reference are different. It is never possible to identify two entities which have different conditions of reference.

On C5:
Human rights are individual rights. As long as we maintain that embryos have human rights a dilemma arises. It is the dilemma of: Who has actually the right to live: Tim or Rüdiger? If Tim has the right to live, then Rüdiger has *necessarily* no right to live. That means in no possible world would Rüdiger have the right to live, even in such worlds he was born. And vice-versa. Then the theory of human rights collapses because this result contradicts the equality of human rights for everyone. And the right to live is a basic human right.

But this collapse only occurs as long as we maintain that embryos have human rights. If the embryo has the right to live then we must ask: Who exactly has this right: Tim or Rüdiger? Otherwise the right to live would not be an individual right. Thus I would reject the claim that an embryo has an individual human right to live. Therefore, it is never a matter of rights whether Tim or Rüdiger should live, but simply a matter of chance. Since individual humans arise by birth, then there is no further conflict of rights between Rüdiger and Tim, because they could never live together in order to establish a dilemma.

Now I turn to metaphysical objections against my result in general. One could object that the souls of Tim and Rüdiger are identical. And so no dilemma arises. But why should the soul be identical since both are necessarily different persons? Even under biological circumstances the same embryo would have different phenotypes if it was implanted in Inge's or in Anke's womb. Biology has proven that there is no genetic determinism as to how the human body develops. Rather it develops in interaction between genome and environment. So we could conclude that the embryo in Inge's womb would develop in a different phenotype than if it was implanted in Anke's womb.

So what is meant by "soul" in the objection here? Is the soul entirely different from the body, entirely different from the biological development? This would lead to a dualism, the consequence of which would be that only God could see that Tim and Rüdiger are identical whereas under logical conditions they are necessarily different. It seems to me that this dualism ignores logical truths. It contradicts the logical structure of reference. This could only be a reality beyond our mundane reality. But we would not have any access to such a reality beyond our own. So it should not have ethical relevance.

3 The power of feelings in bioethics

I assume that my discussion about the ontological state of the embryo evokes feelings. Even if you accept my argument that embryos do not deserve human rights, you will feel uncomfortable in thinking so. I assume that no one will say: "Ok. Embryos do not deserve human rights. So let them use for medical purposes! Let us destroy them for good reasons". We will remain reluctant to say that, because we feel differently.

But we should bear in mind that there are scientists who think so. They have no scruples to destroy embryos. They may not only accept my arguments. But they will also practice killing embryos. Do they have no feelings at all by doing this? Or could we evoke such feelings in discussion?

In my opinion the feelings are actually the point in many bio-ethical problems. Many bio-ethical practices make something *with us*. They touch our conscience. They threaten *our* moral balance. Something happens with us if embryos are sacrificed for medical improvement. And I assume that this moral attitude is missing when research with embryos is routine. In this chapter, I will draw out the emotional aspect of bio-ethical problems.

First I will develop my argument about the structure of such feelings. Then I would like to show in which bio-ethical context this structure becomes apparent. And finally I will discuss the case how we should estimate the biotechnological practice without scruples.

3.1 Feeling, Wholeness and Nature

In my opinion, we are confronted with religious feelings. This does not mean that we feel harmed in our theological doctrines. Rather what I call religion is an immediate human feeling even if one has no clear consciousness concerning it. Thus our problem is not that our doctrine is concerned when we say: "Life is a gift of God. Also the embryo shows God's will to give life. And so we have no right to destroy life, even not the embryo's life". This might be, but behind that statement there are feelings. I think that such a statement interprets our religious feelings.

As a rational statement one could be skeptical whether this is the case. Is God really the giver of life? Does he really want that every embryo should live? What's about embryos produced by IVF? What happens with cloned embryos? Should they all live? Do we have the duty to implant them in a womb so that they could be born? Medical reasons would seem to contradict this statement, because we have not still found a way how cloned embryos could survive even in the womb. And also the birth rate of IVF-embryos is significantly lower than that of eggs fertilised by sexual intercourse. So I think that in a rational perspective, the statement about God as the giver of life is disagreeable, not only for atheists but also for Christians.

Our problem is another. We have to cope with feelings, with an immediate entry to the reality. Our feelings are a warning system. Without feelings we could not apply our rational judgments in the world. And without feelings we could not be moved to make rational judgments.

Our feelings are a warning system. They sense that there is something different; something is remarkable; something makes us disturbed. Feelings need not be concrete. They may not know what they address. But, despite this they happen. They help us to reflect upon what happens. Such feelings are more immediate than our rational capacities and more immediate than our doctrines, even our theological ones. Despite this they could have to do with religious "facts" or religious contents. And this is what I would like to draw out here.

In 2004, the German philosopher Ludwig Siep published an ethics of nature in which he refers to several bio-ethical issues[11]. Therefore he develops a theory of understanding what happens with us when we regard several bio-ethical practices. He discovers a special kind of feelings which are connected with such bio-

---

11   L. Siep: Konkrete Ethik. Grundlagen der Natur- und Kulturethik; Frankfurt a.M. 2004.

ethical innovations. This kind of feelings has a character of wholeness. That means one does not feel something special, something significant, but one feels an atmosphere of an unspecified whole. And this is the feeling of nature.

Thus nature becomes defined as an unspecified whole[12]. It is not a specific kind of entities – plants or untouched landscapes – which is subsumed in nature. But it is an atmosphere in which everything of the unspecified whole is entailed without being specified. Thus nature becomes an emotional term. It is the term of felt wholeness before one describes differently what nature entails in detail. Nature is not a rational thinking of details but an emotional sensing of the whole.

The advantage of this understanding of nature is that the phenomenon of nature becomes saved in face of the biotechnological revolution. By technological innovations it seems that nothing remains what it is. Could we say that trees are natural since landscapes are built like architecture? Could we say that plants are natural since industrial emissions influence the genetic mutations of plants? Could we say that plants are natural since farmers use genetic manipulations for farming? As long as we understand specific entities as natural, the term nature becomes increasingly unrealistic.

However, we feel what nature is. We look onto a landscape and perceive nature. Perhaps this look entails cultural buildings, for instance houses or products of industrial engineering. Despite this, we would say that it is a view of nature. What we see there are not specific entities, and we do not distinguish what is nature and what is not within our perception. But what we see is nature on the whole. Nature in the perception includes cultural entities because nature is not a rational concept but an emotional impression. It is an impression of an unspecified wide panorama.

It was romantic philosophy of the 19th century which has shown that such impressions of wholeness play a necessary role for experience and knowledge at all. Without having an impression about wholeness we could not begin to give

---

12   L. Siep: Konkrete Ethik; Ibid., 214.

an order to our experiences. Without having an impression of wholeness we might see individual entities, but we could not think them in a context of being in the world. The world is nothing which could be perceived. The term world is a rational concept which could never be experienced rationally. Everything we perceive, we perceive within the world, but we could never perceive the world itself. Thus we have no rational reason for knowing whether we indeed live in such a continuum like the world. We could only imagine to live in such a unity like world but it is nothing we could prove by experience.

Unless we feel the world! The continuum in which everything happens we experience is a felt continuum. The American philosopher Charles Peirce called it a faith in a universal continuum. And such feeling of the whole has to do with nature. Thus nature plays a necessary role for experience and knowledge.

Ludwig Siep seems to reinforce this important role of nature for our thinking. And he points out the practical role of nature, the ethical meaning of nature for our concepts of thinking and acting. Nature becomes a feeling which directs not only to an unspecified wide panorama. Nature also becomes a feeling which directs to ourselves[13]. It establishes our self-understanding. Nature does something to us. This is simply because we are included in that wide panorama. The perceiver is a part of the perceived unspecified wide panorama. The perceiver is a part because of two reasons: Firstly nature is a feeling which is based on the perceiver. Secondly the perceived panorama is unspecified and wide so that there is no competence of specifying or making differences between subject and object. In nature, everything coalesces to a whole.

I mentioned that the feeling of nature has practical consequences and even ethical consequences. This has to do with the kind of feeling. By which feelings do we sense nature? What are the emotions which are associated with nature? It is a kind of agreement, of relationship, of respect. Nature is an accepting feeling. I perceive a panorama in which industrial buildings may be included and perhaps a plane flying overhead. Despite this, it is nature. Everything is in har-

---

13 L. Siep: Konkrete Ethik; Ibid., 218.

mony, and nothing is disturbed because everything is included in an unspecified, wide panorama.

According to Siep, it seems to me we could define nature as follows: *Nature is a feeling entry to things without differentiating them sharply; rather that feeling accepts their interaction in a perceived unspecified wide panorama; and this kind of acceptance has no specified reasons.* The latter remark is significant for nature: We do not accept nature for any instrumental reasons. We do not accept nature for ecological reasons or because we want to save the environment. Nature has no political motive and even no moral move. Nature is an aesthetic term: We accept nature for no specified reasons. That does not mean that we have no reasons at all. To have no reasons would mean to have no engagement. But whoever perceives nature is engaged in nature because he himself is a part of it. However, there are no sharply distinguished reasons why nature feels nice like harmony.

I assume that present scruples in bio-ethical issues have to do with feeling nature. Feeling nature will be disturbed when embryos are killed for medical improvement. Natural feelings are disturbed when genetic enhancement of the human genome becomes possible. Natural feelings are disturbed by Neuro-Enhancement when brain-capacities are improved by chips or micro-technology. The wholeness falls behind precise planning of what life should be. In my opinion, the crucial point is that our natural feelings are confronted with such technological innovations. The ethical problem is not the moral state of embryos; the problem is not the ontological state of genetically enhanced infants. And the problem is not the moral state of people with neuro-enhanced capacities. But the problem has to do with us, with our capacity to integrate our sensations in a whole by accepting it without special reasons.

By biotechnological innovations, our capacity to integrate our sensations in a whole becomes deflated. And this has serious consequences for our knowledge and our practice. But before I will illustrate this hypothesis, I am going to add a further remark.

Beyond Ludwig Siep, I would like to interpret the term "nature" more precisely. In my opinion nature has a religious dimension. Nature is a phenomenon which *befalls* us. We could not produce the impression of nature. We could not create that impression. But we are dependent on the fact that nature befalls us. Surely we could plan a hiking trip of through fields, mountains and forests. But the capacity to experience nature is uncontrollable. It must befall us or we have no natural feelings. I have already mentioned that nature has nothing to do with a special kind of entities like trees or plants or seasides. But it is the feeling of infusion of all what we perceive with the perceiver. And this event of infusion is uncontrollable. We could be open-minded for the befalling of nature, but we are not capable of evoke the befalling. Nature essentially reveals. It is a kind of revelation.

It is far from clear whether this revelation is a Divine one. It might be but even if it is not it seems to be a religious feeling. The term "religious" means: It is out of human control although it has to do with our self-understanding. Thus I would like to conclude that human self-understanding depends on religious grounds, that is at least the impression of nature. This essential feeling for self-understanding and of understanding the reality as a whole is a religious one because it befalls us. And bio-ethical innovations threaten the balance of self-understanding. It devalues religion and thus it changes our mode of self-understanding. And this is the reason why we may feel scruples when we are confronted with bio-ethical innovations. The scruple is the warning system that something could happen with our self-understanding.

To sum up: Feelings are evoked by biotechnological innovations. And they have a typical structure: They direct to nature as an unspecified whole we accept without specific reasons. This natural impression also directs to our self-understanding. And it has a religious basis because nature could only befall us. The scruple in bio-ethical issues seems to suggest that nature is in danger: The religious background could be blanket by biotechnological innovations: Befalling of nature could be weakened. Then nature could become less valuable or hardly experienced. And finally, our self-understanding is involved by such changes of our capacities to realize experiences. This is my hypothesis, which I would like to prove now.

3.2 How does biotechnology influence our feelings

Let us compare the experience of spontaneous pregnancy with IVF. When a woman becomes pregnant this is a very exciting moment. Usually it is a glorifying moment in which everything gets the shine of having a child. It might be that the parents begin to adapt their actual plans and to revise them for the baby. But the baby is welcomed. And it is welcomed in a world which is okay. It seems to me that in such cases there are elements of nature in the parents' feelings, at least the acceptance of the world without specified reasons. Otherwise the baby could not be welcomed.

I concede that many couples fall into crisis when they realise that they are becoming parents. But as long as they accept the childbearing itself, they accept the pregnancy in an unspecified wide whole in which everything is accepted without specified reasons, their anxiety included. Also, the anxiety would be a part of nature then.

This changes through IVF. IVF makes it possible to select embryos by medical criteria. Genetic screening of the embryo's genome, pre-implantation genetic diagnosis leaves room for the choice for or against the implantation. Selection of embryos becomes possible. In almost all countries, usually, IVF fertilizes more embryos than the parents wish to have, simply because there is a high rate of risks by IVF. So even if a society forbids the practice of pre-implantation, genetic diagnosis, selection could happen simply because the woman could refuse to implant all fertilized eggs. Thus the IVF-practice pushes the impression of nature away. The embryo is not welcomed without specified reasons; the embryo is not welcomed in the environment of a unspecified wide whole. But the embryo is only welcomed if it fulfills specific conditions. In this context life loses its glory, its universal shine.

Biotechnological innovations change the natural character of life to more precise criteria what life should be. Instead of nature existing as a revelation, life-giving becomes a matter of human responsibility. Another example: Instead of ideas occuring in the mind, neuro-enhancement improves the brain in order to control mind-actions.

Surely the question will arise: What is better? Nature or responsibility? Befalling revelations or moral decisions about the life chances of beginning life? Is selection wrong in itself? The crucial point is that by this question we seem to compare two dimensions of thinking which are not comparable. Responsibility is a moral dimension, but nature is an aesthetic term. Therefore, we have no way to compare biotechnology with natural impressions directly. Despite this, nature plays an important role for moral self-development. Nature is a pre-moral condition of a moral self-understanding. We need natural impressions in order to integrate our experiences in the whole we perceive as a continuum for our life. Nature is an essential impression for theoretical and practical orientation. Thus we have the responsibility to protect the possibility of befalling nature-impressions.

It might be that IVF will not destroy natural impressions at all. IVF might support medical improvement in only some references. As long as nature remains a possible experience there is no complete threat to human self-understanding. Technology might not contradict nature at all but only in some references. Then it might be responsible to establish these techniques as long as there are human resources to experience nature. It seems to me that IVF could pass this test.

In a publication of 2005, I have argued in a similar way about cloning of embryonic cells[14]. I think that in some cases it could be useful to confront a technological routine with symbolism of nature. Suppose an institute of research of

---

14 L. Ohly: Das Erschrecken vor sich selbst; Ibid., 165–167.

embryonic cells. Suppose every embryo which is killed by research would get a name on a plate within that institute like a tombstone. This would be a symbolism for the natural acceptance of embryos without differentiating. And such a symbolism would disturb the routine of killing embryos.

Such a symbolism is not ethical, because it has nothing to do with reasoning. Rather the rational objections against calling embryos by a human name seem more persuasive as we have seen in Kripke. Thus such a symbolism would not be ethical. It is aesthetic. It is an aesthetic provocation to scientific practice. But we could have good moral reasons for that aesthetic demonstration. One good moral reason is to compensate the natural perspective against scientific threat of specification, of differentiation and of selection.

I have suggested a rule how to justify biotechnological innovations. This rule seems to be paradoxical. It is the following: As long as scruples arise through biotechnological improvements, we are sensitive enough to feel natural impressions. And as long we do, the practices could be tolerated. Sometimes such techniques should be confronted with symbolism of nature. As long as this symbolism could correct the capacities of perceiving the world as a whole, technology could be tolerated. But since natural impressions get lost by the suggestion of a biotechnological totalitarianism, the claim of biotechnology is too strong and ethically wrong. Essential pre-moral conditions for self-understanding and for integrating experience in a whole would be harmed.

3.3 Why do scientists have no scruples?

Is it possible to lose the natural background for thinking and experiencing? And are scientists really in that situation? If the answer is yes, they would have a psychological burden because no one could think and experience without assuming a continuum of the whole which is neither concretely thinkable nor perceptible. Indeed, scientists regard individual phenomena, they are concerned to single facts and not for the whole. We have no universal scientific theory about the world. There is no cosmological framework in which all single facts could be integrated. Too many paradigms contradict each other, too many theories could only explain parts of the reality without having a link to the environmental theories. Thus it might be that a concept of the whole contradicts the kind of scientific thinking today.

On the other hand, scientists live their lives like other people do. They have no problem in daily life's orientation. I do not think that scientists have psychological burdens. What I think is that scientists actually have a natural orientation but this natural link is hidden by other routines of their scientific practice. The scientific practice leaves no room for cosmological frameworks, neither theoretical nor emotional.

In Germany, I had a colleague, Thomas Wabel. Wabel has proposed that scientists also have similar intuitions to people who have scruples with biotechnological innovations.[15] According to Wabel, both perspectives are of the same nature. People who have scruples could not claim to be closer to reality than scientists who have no scruples. And scruples are not morally better than intuition or more immediate than scientific intuition, in which scruples are missing.

Therefore, Wabel draws out a theory how moral intuitions evolve. No moral intuition is eternal, nor is an intuition based on fixed anthropological constants forever. Otherwise there would not be plural intuitions and a discourse about moral intuition. Wabel thinks that there is no universal rationality shared by everyone. Rather, intuitionis the result of interpretation and vice-versa: Interpretation is the result of intuition. There is a circle between intuition and interpretation. Both are dependent upon each other. It is impossible to find the origin of intuition, especially to find them in innate moral skills. Wabel describes an evolution of moral intuition. The result of his description is that the

---

15 Th. Wabel: Intuition und Interpretation. Zu Grundlagen der Konsensfindung in ethisch relevanten Fragen des Zellkerntransfers (SCNT); in: P. Dabrock, J. Ried (Ed.): Therapeutisches Klonen als Herausforderung für die Statusbestimmung des menschlichen Embryos; Paderborn 2005, 331–356.

intuition of scientists is the same as scruples against biotechnological innovations. There is no moral fault to think that an embryo in its earliest stages is only a bundle of cells as scientists think.

It seems to me that Wabel's theory may be true in principle. But it fails to show whether the different intuitions are really based on the same original level. It is one to say that every intuition is based on interpretation, on personal experience and perhaps education. But it is something different to say that every intuition has the same original level. It seems to me that scruples against biotechnological innovations are founded more deeply than the intuition of scientists. As I already mentioned, scientists too need a natural impression in order to integrate their experiences in a whole. Therefore, they need the dimension which would actually lead to scruples. But they override this dimension by developing further intuitions.

In other words: Scientific intuitions are more contingent, they are more arbitrary than the moral intuitions of scruples against biotechnology. Scruples are based on a necessary feeling of nature which is necessary for human orientation in the world. So I disagree to Wabel's suggestion that all intuitions are of the same value or the same moral weight. In my opinion, this is a prejudice Wabel cannot prove by his method. Rather it must be the result of analyzing intuitions which one has the more deeply founded moral ground.

To sum up: Scientists have moral intuitions which differ from those who have scruples against biotechnological innovations. These differences are still to be evaluated. The pure fact that differences in intuition exist does not lead to the conclusion that all of them have the same moral weight. Wabel takes the so-called "natural fallacy": He concludes a prescriptive statement from a descriptive statement. He concludes that all intuitions are right from the empirical fact that there are plural intuitions.

4 Human reproductive cloning and germ line therapy

This chapter is related to the previous chapter. I speak about our feelings in order to evaluate the scenario of genetic enhancement. Usually ethicists refer to potential human beings who could be created by such forms of genetic enhancement. They speak about "psychological burdens" of the persons so created. They stress that genetic enhancement hurts human dignity. In my opinion, this is not the right strategy to analyze scruples against these practices. Rather, the way of arguing should be the same as before: *We* are harmed since human beings are cloned. *We* are harmed if other persons are created by germ line therapy.

4.1 Reproductive Cloning

First of all: What do these kinds of enhancement mean? Reproductive cloning means a technique to produce a genetic identical twin of a person. The purpose of reproductive cloning is to bear the cloned being. It is a kind of human reproduction. The core idea is the following: One takes an unfertilized egg and extracts the nucleus of the egg, that is the matter which entails the genetic material of the egg. Then one incorporates the nucleus of another cell into that egg. This other cell derives from a person, perhaps from an adult person. It is important to reprogram that nucleus so that it becomes totipotent again. Afterwards, one originates an electronic impulse so that the egg divides in further cells: Other cells gestate and the process of embryonic development begins.

By this case of reproductive cloning – if it is successful – the born infant will be genetically identical with the donor of the nucleus. They will be twins although they have entirely different ages. The brother (or sister) could be already an adult.

One usual objection against reproductive cloning stresses that this technique is entirely ineffective for mammals. 1997 the cloned sheep Dolly was born. Unfortunately one has tried to clone a sheep 277 times before Dolly was born. This is a miserable test rate. Moreover, Dolly died much earlier than it is usually expected for sheep. One finds out that the faster process of maturing was genetically based. The conclusion is that reproductive cloning is ineffective and undesirable because of the disadvantageous aging.

To improve the rate and the aging for humans one would need to test with human clones. Thus human fetuses would become objects of experiments which could die during these experiments. And the first human clones are expected to have a bad mortality rate. This is a kind of testing humans which seems to hurt their human dignity.

I am skeptical whether these objections are valid. The philosopher and ethicist Derek Parfit has developed an interesting analogy which is often quoted by the proponents of cloning. Suppose there is a woman who should not become pregnant now because of an actual sickness. Physicians told her not to become pregnant because otherwise the infant could get a handicap. But the woman ignores the advice and gets pregnant. The consequence is that indeed the child is handicapped. After growing up the child accuses his mother to have ignored the physician's advice. But the mother answers: "If I had not ignored the physician's advice you would not be here. The only way for you to live was to live with your handicap".

Proponents of human cloning argue in the same way. It might be that human clones will have a bad mortality rate. But otherwise clones could never live. Should not it be better to let them live even if their mortality rate is high? Or do we hurt their rights if we let them live at all?

It depends what is meant by human dignity. I will discuss that issue later on in another chapter. But Parfit's argument is "prima facie" a good answer against the objections of reproductive cloning.

There is another class of objections against human reproductive cloning which seem to be more interesting. They focus on the problem of self-understanding and self-development – issues I have already mentioned when I was talking about the structure of natural feelings. Many authors object that a twin was only constructed because she should fulfill hopes in her who is identical with another. The hopes in a cloned twin are different from hopes in another child. What she should learn, what abilities she will have, which career she could achieve – these hopes become more concrete in clones than in other infants. For instance the adult twin who is the donor of the nucleus has been a wonderful and famous pianist. Suppose after an accident he has now lost his right arm. Now the hope is that the genetic twin will have the same skill. After birth his parents – note: the father will also be his twin brother – will educate their child in order to become a nice pianist like the father.

We all know that often times parents hope in some development of their children, and they influence their childrens' education so that the children could reach the aim their parents have imagined. But the point is: Clones suggest to *be* like another who has already existed and who has already performed his skills. The suggestion that the clone will be what his parents hope to be will increase. The opponents of human cloning argue that this suggestion harmed their "right of ignorance". Humans have the right not to know which genetic capacities they have (Hans Jonas). And secondly they have the right to an open future (Joel Feinberg). Both rights would be harmed by reproductive cloning. The clone is suggested to be another person whose life history has already gone. So the future of the clone seems to be fixed by the hopes and expectations of his social environment.

Scientists might object that this is a kind of genetic determination. Indeed a genome develops not automatically equally in all circumstances. Rather it develops in interaction with its biological environment. The mother of the clone is not the same as the mother of the nucleus donor. The families are different. The behavior of the mother is different and so on. So we could conclude that the clone will not be the same one as his twin. And all hopes to fix his future are only fantasies.

However such fantasies are the motivation to clone people. One would like to clone people because one desires to create twins of concrete persons whose history has already passed – at least partially. Thus the fantasy of cloning is a strong suggestion of reassuring *that* life of the older twin if we use that practice at all.

This leads us to the question: What could be the reasons to perform reproductive cloning? In discussions one finds mainly the following four motives[16]:

1) Reproductive cloning could be a helpful measure when a couple is infertile. By cloning, it would become possible to have a child who is biologically linked to at least one part of the parents. Moreover, reproductive cloning is an alternative to hormone therapies which should improve the fertility of the couple. Such therapies are expensive and risky.

2) Gay couples could get a child which is biologically linked to at least one part of them.

3) If a child of a family has died and if there is genetic matter frozen from the child then it becomes possible to reproduce a clone with the same genetic make up. The couple could get a twin of that child. Perhaps this could be a way to help the couple after the death of its child. The grief could be conciliated. It is apparent that the hope to get the dead child back is an illusion. But proponents of reproductive cloning leave it open whether that hope is a realistic chance for family support despite this. Especially liberal proponents argue in that way; that couples have the right for constructing hopes even if they seem to be illusory.

4) Finally a science-fiction-example: A global catastrophe could happen. Many people could die and only some people could survive because they have a

---

16 D. Birnbacher: Aussichten eines Klons; in: J.S. Ach/G. Brudermüller/Christa Rautenberg (Ed.): Hello Dolly? Über das Klonen; Frankfurt a.M. 1998, 46–71, 49f.; D.W. Brock: Cloning Human Beings: An Assessment of the Ethical Issues Pro and Con; in: M. C. Nussbaum/C.R.Sunstein (Ed.): Clones and Clones; Facts and Fantasies about Human Cloning; New York/London 1999, 141–164, 146; D.W.Brock/N. Daniels: Why Not The Best? in: A.Buchanan u.a.: From Chance to Choice. Genetics and Justice; Cambridge 2000, 156–203, 200; W.N. Eskridge Jr./E. Stein: Queer Clones; in: M. C. Nussbaum/C.R.Sunstein (Ed.): Clones and Clones; Facts and Fantasies about Human Cloning; New York/London 1999, 95–113, 95, 97; H. Jonas: Technik, Medizin und Ethik. Praxis des Prinzips Verantwortung; Frankfurt a.M. 1987, 172.

supportable genetic make-up. Should not it be socially important to improve a human mankind with similar genetic make-up? By that measure the survival of human mankind could be saved better.

Surely, it seems to be that every example provokes objections. Does infertility justify the innovation of a risky technology? Do have gays the right to have own children? Does the wish to get a child back hurt the right of ignorance and the right to an open future? And finally: Could an extreme exception like a fictional global catastrophe justify establishing a technology which should become routine? Could be a technological routine justified by such an extreme exception? – These are serious objections. And we should enter a discussion in order to decide how strong the arguments are. But if we do so it becomes apparent that by this way it is not a question of principle but of waging pros and cons. By this way it seems to me that we have not to do with principal objections against reproductive cloning which would justify a general ban of that technology.

4.2 Germ line therapy

Germ line therapy is a medical intervention to the genome of sperms or unfertilized eggs or even of an embryo. By germ line therapy the genetic make-up of an individual could be enhanced. For instance if the genome shows a defect which would lead to a serious sickness after birth, the germ line therapy could repair the defect. This would be a kind of healing or better of preventing harm. Germ line therapy could also improve genetically based potentialities for actualizing human skills. For instance it might be possible to improve the conditions for muscle building, brain capacities and lifespan.

One problem of germ line therapy is that it does not only address the patient. Rather it addresses everyone who is connected within the genetic link to that intervention. The genetic intervention does not only refer to the embryo but to the whole offspring of it. This changes the meaning of medical intervention, the paradigm of the relationship between physician and patient. The physician would treat patients who are not still born and who may live decades or centuries later on. The paradigm of medical treating in context of the informed consent of the patient becomes questionable simply because they are not still all patients in present who could consent to the treatment. Thus a modern liberal justification of medical treatment collapses.

By this way, the physician would not entirely take responsibility for all the consequences of his treatment. He could not be accused at a court since further generations could prove that he has been mistaken. In this situation, the meaning of responsibility changes so that it could become questionable whether responsibility is still maintained by germ line therapy. It seems that germ line therapy could never be a responsible decision.

Moreover, the line between therapy and enhancement dissolves. Even if one justifies germ line therapy as a medical treatment, it is doubtful whether one is able to justify enhancements in the same way. Would improving capabilities be within the essence of medical treatment? Is it the task of medicine to improve human capabilities? Suppose it is not. Does it then generally follow that it is unjustified to enhance human capabilities? It seems to me that the difference between therapy and enhancement is not a sufficient reason to reject germ line therapy. But it shows how medical self-understanding will change through that technology.

There is a third issue which makes the problematic difference between therapy and enhancement more serious. There are good reasons to think that the subject of therapy is not the same as before. Rather, the therapy replaces the patient. Reaffirming Kripke's argument that the original conditions for referring someone are rigid; they are necessary conditions for referring to and identifying the same subject. I have already mentioned in my argument that one and the same embryo does not necessarily mean referring to one and the same individual human (the example of Tim and Rüdiger). Now I argue that germ line therapy will not only change something within the embryo. But the embryo itself will be replaced. After germ line therapy another embryo is derived from the first.

A simple example is the following: Suppose a germ line therapy will change the sex of the embryo. I guess we all agree that a woman is not the same person as a man. So a female embryo could not be the same one as a male embryo. The conclusion is that changing the sex by germ line therapy would not treat the embryo but replace the embryo by another one.

One could object that this intervention is a fundamental one which does not only replace single genes but a whole set of chromosomes. But on the other hand the genome functions systematically. That means if only one element is changed the whole system runs differently. This is a biological fact concerning the function of the genetic system. So it seems to me that my example about the sex change is only gradually different from changing single genes. It is not a difference in principle.

However, one could further object that also during the lifespan our genome changes incessantly. But no one would say that I am replaced by another person if a tiny single genetic mutation arises. I concede that. But I think this objection supports Kripke's flexible criterion on how we refer to beings. I could not be re-identified by referring to my genetic make-up. The conditions of me are the ones which correspond to the original situation in which I was referred to the first time. One property of this original situation was a special genetic make-up which may be different to my present genetic make-up but which is historically connected as a continuum. Moreover, one need not to refer to my original make-up when I was born whenever one refers to me. Therefore, it seems sufficient that I have had a special genetic make-up in the original situation which is historically linked up to my present one in a continuum.

In comparison, the genetic make-up of the embryo is essential for referring to it simply because otherwise germ line therapy would not function if one ignores the genetic make-up of the embryo. One must concede that a genetic intervention changes the original conditions of the genetic system which is the condition for referring to the embryo. There is a difference if a genetic system changes by continuous interaction with its biological environment or if an abrupt intervention from outside replaces the rules of interacting between genome and environment. So I tend to say that a germ line intervention does not change something within the embryo but it replaces the embryo by another. It only uses the same embryonic matter.

To sum up then: Germ line therapy differs from reproductive cloning because it is a real medical intervention for healing. But it also confuses typical relationships:

1) between physician and patient;
2) between physician's responsibility and the wide horizon of the medical treatment the physician cannot foresee. Therefore he cannot identify his area of responsibility.
3) between treating and enhancing;
4) between treating and replacing the subject of treatment.

4.3 An anthropological framework to evaluate cloning and germ line therapy

In 2001 and 2007, I have developed a phenomenological theory for evaluating these two technologies[17]. Phenomenology is a philosophical theory of experience which searches for evident structures within a perception. One of the first philosophers to have developed the phenomenological method is the German Edmund Husserl (1859–1938). Nowadays, the phenomenological method is recognized worldwide as a philosophical discipline. Husserl's interest was to develop a theory of knowledge which transcends the opposition between objectivity and subjectivity, or between realism and idealism. Traditional philosophical concepts either describe the essence of entities as if it was evident that they are real (this is realism, objectivity). Or they describe all entities as

---

17 L. Ohly: Der gentechnische Mensch von morgen und die Skrupel von heute. Menschliche Leibkonstitution und Selbstwerdung in den prinzipiellen Einwänden an Keimbahntherapie und reproduktivem Klonen; Stuttgart 2008.

conceptual - that means they depend on our conceptions about the world (this is idealism, subjectivity).

For example: Is it right to say "There is a table"? Is it evident that there is a table? Or is it possible that we all only dream that there is a table? Is it only an invention of our mind that there is a table? Moreover it might be that even if there is something, we could not evidently conclude that this something is a table. Maybe we only see that something as if it is a table. Our perception or our imagination of that something is based on our conceptions of the world. But then we could never be certain what there really is and what happens in reality because we have no evident access to the reality. We have only a conceptual-based perspective to something we call reality. The conclusion is that the question what reality is depends on our conceptions on what we mean by reality. This is the subjective opposition to realism.

Now Husserl has developed the phenomenological method in order to override the gap between subjectivity and objectivity. He would agree: We could never be certain that there is a table. But we could stress that we perceive a table there. At least for me it is evident that I perceive a table there. And you will agree to the following statement: If I perceive a table there, then it is evident for me. But this kind of evidence is not only a subjective one. It is also evident to you although you need not perceive a table there and you could never know if I really perceive a table there. Despite this, it is true that I evidently perceive a table there, if I perceive one at all. This kind of truth is neither objective nor subjective. It is phenomenological. Thus phenomenology deals with evidence in perceptions and imaginations. Moreover, this method never decides if the mental state corresponds to something in the world.

For my ethical interest, it is sufficient to describe the evident conditions of perceptions, feelings and conceptions. Moreover, I think that there are evident reasons to assume a transcendent perspective one could identify with a Divine one. Husserl himself has refrained from religious perspectives in his method. But there are many other thinkers who have embedded religious and theological knowledge within phenomenology, e.g. Emanuel Levinas and Jean-Louis

Marion – who seem to me the best philosophers who have combined phenomenology with theology.

These remarks are only preliminaries. Now I would like to point out my phenomenological theory, which seems not to have to do anything with bioethics at the first glance. I beg for patience. It is simply a theoretical framework in order to show what genetics will change in our fundamental conditions of self-understanding and self-awareness. Before I draw out my theoretical framework I would like to give a short summary of the result of my theory. I think that germ line therapy and human reproductive cloning ignore natural feelings again. They suggest an understanding about human mankind which seems to be entirely objective. They push away the subjective element of self-awareness. Both methods of genetics suggest that only an objective perspective of human mankind is relevant for human self-consciousness. Complicated combinations of subjective and objective perceptions, which are typical for feelings for nature are left out. The suggestion is a perspective of the third person, and it seems to eliminate the perspective of the first person, which is phenomenologically essential.

If this suggestion was right, it would change human self-awareness entirely. It seems to me that this suggestion is wrong. It is impossible to refrain from natural feelings and to transform subjective knowledge about humans to objective representations. But as a suggestion, it is very powerful and will dominate human self-understanding even if natural feelings are also entailed in human self-awareness. But it will not be a conscious suggestion, because of the strong domination of genetics.

Now I present the phenomenological theory. We are corporal subjects. We have an inner life but also an apparent representation of ourselves which is our body. This is a tension between inner and outer life. And this tension is based on inverse (opposite) bodies of evidence. Some facts of the outer life are evident whereas they are not evident for the inner life, and vice-versa. A case in point is the following example: I know evidently that something hurts in my self. I have evidently a subjective feeling of pain. My partner could not evidently know if I am in pain He could only guess or assume that something hurts in my self.

Vice-versa: my partner knows evidently whether there is a mark on my face. He simply looks at my face and knows it. There might remain the philosophical problem whether my face really exists or whether it is only imagined. But also, if it is imagined, it is evidently clear that my partner could verify directly and evidently whether there is a mark on my face. In opposition I cannot directly verify whether there is a mark on my face.

I call this phenomenon the inversion of perceptions. Although I have an immediate self-awareness, I could not know everything about me. I could not know everything about my outer life. In the face of my outer life I could err. On the other hand, my self-awareness makes it impossible to err in the face of my inner life. It is evidently true if something hurts. The point is that this inversion applies to my relationship to others. Everything I could not evidently know about myself I know evidently about others. And everything I know evidently about myself I could not know evidently about others. In comparison to others, there is generally an asymmetric advantage of knowledge.

The point is that I need the perspective of others so that my knowledge about myself could increase. And vice-versa: others need my knowledge about myself so that their knowledge about myself could increase. But since we borrow the knowledge from each other for completing knowledge about ourselves, we lose evidence about us. It might be that I have a mark on my face. Others could tell me that this is the case. But I cannot still know evidently if this is true even if they tell me. For instance they could lie. Although I could verify their information by a mirror, I will accept the image on the mirror as an image of myself. But the image of myself is not identical with me. An image is not the same as me. So I perceive myself as an other, and so I lose the evidence of myself in perceiving an image of myself. I only turn to another perspective in which I am no longer what I seem to be.

Note that I argue phenomenologically. I argue by analyzing perceptions. In daily life we identify the image on the mirror with ourselves. In daily life, we simply trust in our perceptions and what we are educated to identify. But this does not mean that the truth of such trust is based on evident experience. In daily life trust seems to be enough usually. But trust is epistemologically not as strong as evidence. The phenomenological question is: Is there a dimension within the perception which gives evidence to me that my evidence about me and the others' evidence about me refer evidently to myself?

This sounds complicated, so I will re-iterate what I mean. Others could evidently know something about my outer life. I know evidently about my inner life. But where is the perspective which integrates both bodies of evidence? The others could not because they could not jump in my perspective to feel my inner life. And I could not because I could not jump in the perspective of others in order to perceive my outer life as *my* outer life. Whenever I perceive my face in a mirror I actually perceive only a face. But I do not evidently perceive the "My" of my face. From an external perspective, I cannot perceive my subjectivity. But as long as I am aware of my subjectivity I could not evidently identify my subjectivity with information of an outer life. That is the crucial point in my question: Is there a dimension within the perception which gives evidence to me that my evidence about me and the others' evidence about me refer demonstrably to myself?

The answer is that such a perspective is missing: one which integrates my outer life's information about me with my inner life's information about me. At least there is no perspective which makes the integration evident. I could only trust that the outer perspective about me really represents myself. And others could only trust that I really have the inner life's evidence, which I tell them I have. And this kind of trust is a "blind" trust. The integration of inverse information is not based on perception, it is not based on evidence but on trust.

Despite that, we need an integration of both perspectives simply because we have both sides. We have an inner life and we represent ourselves in an outer life which is immediately perceivable by others but not by ourselves. So we need trust. This kind of trust is the basic precondition for living in unity. The point is that we must fulfill that basic condition generally. It would be a psychological burden not to live in unity. But we need not decide to trust in a kind of higher perspective, no human could enter. Rather we do it simply, without reason or decision. We trust "immediately". Trusting in that third perspective is as immediate as both perspectives of evidence[18].

I interpret this trust in a third perspective as a religious faith. God is the third perspective which integrates knowledge to an omniscient entity. This might not occur consciously. Not everyone needs to become a religious believer. But - unavoidably - every person needs the trust in that transcendent perspective, the only all-knowing perspective. So I tend to say that religious feelings are anthropologically necessary.

For our ethical evaluation of genetics, it is noteworthy that trusting in such a third perspective has the practical consequence of constituting human freedom. Humans remain free because they do not know everything about themselves. More important is that humans remain free because they do not know everything about each other. The integration between my inner life and my outer life never falls within one of those perspectives. Neither others could define who I am, nor could I define who I am. If I was able to define myself I would lose my freedom. I would have known what to do before I could decide what to do. Conversely: If others were able to define me I would also lose my freedom. Others would best know how I would decide to act, so I could never decide. Their knowledge would destroy my ability for decision-making. The consequence is: Because no perspective could integrate these two opposite kinds of evidence, I can live an open-minded human existence. I am free from myself and I am free from others. So I can live open-minded.

A short appendix about a Protestant conclusion: The Reformer Martin Luther (1483–1546) has stressed that human freedom got lost in the face of God. There is no human free will in the face of God. My phenomenological theory supports this Protestant conclusion: The trust in the third perspective is the trust in some-

---

18 Sören Kierkegaard's spiritual integration of soul and body has the same connotation as my phenomenological analysis of inner and outer life. According Kierkegaard ‚the spirit is the integration of two opposite dimensions which nevertheless need to unify because otherwise the human would not exist. The difference to me is that Kierkegaard analyses this integration as fear whereas I call it trust. (S. Kierkegaard: Der Begriff Angst, 42ff.).

one who could integrate both opposite (inverse) kinds of evidence in a transcendent knowledge about everyone. As long as God knows everything about us we could not be free in the face of him. Our unfree will is the back side of God's omniscience. (This statement is only theologically noteworthy but only of limited relevance for our issue of genetics.)

4.4 How genetics disturbs human freedom

Remember both existential objections against human cloning: Reproductive cloning would hurt the right of openness and the right of ignorance. My analysis shows how both rights are connected: Ignorance as knowing only partially about myself is a condition of open-minded existence. Thus ignorance is a condition for openness, for freedom. The objections against reproductive cloning address two typical conditions of human life. It seems that the clone is under domination of social constructions. The social environment dominates the self-development of the clone by strong expectations how the clone should behave and how he should develop. The social environment suggests to know everything about the clone because "he" has already lived – in his twin. The twin is the outer perspective and now also becomes the inner perspective of the clone – at least suggestively. The twin seems to be able to define what the clone is or at least what he should be.

We all know that this is only a suggestion. But it is a strong suggestion. And this suggestion must lie behind the motive for reproductive cloning. The examples why one could be a proponent for reproductive cloning all refer to fantasies of being biologically interlinked with the clones. There is the suggestion to be not only familiar with one's own children but also biologically interconnected or almost fused. The fantasy to get one's own deceased children back; the fantasy to get an extraordinary pianist back; even the fantasy of gay couples to have their own children – what does it mean to have one's "own" children? It seems to me that it entails the imagination of possession which is justified by a genetic link. Note that actually there are many frozen embryos; there are other ways to get one's "own" children, for instance by adopting children. Also other medical aids for child-rearing are already possible. But it seems that reproductive cloning fulfills the dream of a biologically linked ownership and a social ownership of what a child should be and what should happen with the child. It is only a suggestion. But without that suggestion one could never be a proponent of reproductive cloning.

Surely, heterosexual couples also have an interest to have their biologically interlinked „own" children. This is usually an innocent desire. A similar ethical problem arises, since one constructs a "right to have one's own children". If there is a right, then the society has the duty to support having own children. Behind that claim, there is the social construction of the "own child" which weakens the openness of the child. Now, in reproductive cloning, the claim is stronger because the child is not only biologically interlinked but the real twin of one family member. The child is not simply like us, it is not familiar with us, but it is me – even if I have had only the desire to have my own child. The only way to get one's own child is - by suggestion - reproducing oneself. This destroys openness even if I had not wished it so.

Phenomenologically, the suggestion is that I give myself for another who becomes myself. I play a double role. It is the role of inner life and of outer life together, in both cases with the suggestion of having evident knowledge about me. My knowledge about my inner life addresses two persons, and also my knowledge about my outer life. Because I know about my inner life the suggestion arises that I also evidently know about the inner life of my twin. And vice-versa: Because I know about my outer life in the face of my twin, the suggestion arises that I also evidently know about my outer life. It is only a suggestion but a strong one and an inevitable one since I argue in favour of reproductive cloning.

It seems to me that the suggestions even increase in germ line therapy. We have determined that replacing one element in the genome changes the whole genetic system. Thus, whoever wants to replace one gene has an entirely new concept of a human in mind. As long as germ line therapy is not epistemologically well proven, such a concept will be missing. But then one could not justify such a treatment of the germ line. But since this technology is sufficiently proven to the extent that one could start to use it, one will have a whole concept about the interaction between genotype and phenotype - that is, between the genetic frame and the actual development of a human. An ethical dilemma arises: Do we want to respect the openness of the development of an embryo? But then we could not justify a treatment of the germ line. Or we do respect it. But then we have a concept of how genotype and phenotype will interact in order to justify the treatment. Thus we would harm the right of openness.

Actually, there is no strict interaction between genotype and phenotype. How the phenotype will develop depends on several factors and not only on the genome. But the suggestion that we are responsible for the development of the manipulated person will necessarily become stronger. We adopt his freedom. And this means a fusion of the other and us. We usually adopt properties the other has and not us. Naturally the other is free, because there is a gap of inverse kinds of evidence. Now we adopt his freedom. A fusion between the inner life and the outer life is suggested.

Phenomenologically germ line therapy has at least three possible ways to harm the right of ignorance and the right of openness:

1. Semantically, the meaning of openness will be reduced. What nature is, which feelings are built in perceiving wholeness in an unspecified wide panorama – this might become meaningless. Naturally, there is an unspecified unity of me and the other integrated in a third perspective of the transcendent. That is the reason why I think that nature is associated with religion. Natural feelings are religious feelings. There is a unity between myself and the other. But this unity is neither integrated in my perspective nor in the perspective of the other. It is constituted by a transcendent perspective of the Divine. Natural feelings correspond to the trust to be integrated in an absolute perspective of the Divine. This will change by germ line therapy. The space of symbolizing unspecific whole-

ness will be reduced. Germ line therapy will decrease the sensitivity for such phenomena.

2. Through germ line therapy the person with a manipulated genome seems to be absolutely explainable. We do not meet him open-minded. But we look onto him with our conceptual expectations we have constructed in germ line therapy. Let me give an example. Suppose a medical diagnosis has shown that a genetic defect could lead to a brain damage. Now some genes would be replaced in order to eliminate that risk. The therapy does not only eliminate a symptom of a patient. But it changes the whole rule of a systemic genome. Therefore a concept is necessary in order to justify that change in the system. And this means that the physicians must consider how the patient will behave after the treatment. In such cases the way of living seems to be determined by the germ line treatment. Not only the physicians but the whole social environment is sensitive to the change of behaviour after the treatment. And everyone will control and verify whether their conceptual expectations will happen.

The suggestion is that we will know him better than he himself. He could only evidently know his inner life, but now it seems that we could also evidently know his inner life because of our concept of him which was the justification for the germ line treatment. It is only a suggestion. But even if the person with a manipulated genome makes apparently free decisions, we will be already there and could tell him that he precisely corresponds to the concept we have constructed about him. It is only suggestive but unavoidably compelling.

By the way: In many articles about this issue one finds the metaphor of "Playing God". Germ line therapy is an expression of playing God. Most of them understand this metaphor as correcting creational conditions of our life. The simple implication behind the Playing God Metaphor is that actually the human genome is good; it is well created by God and needs no correcting. In my opinion this sounds too simple because we all know that many diseases are based on genetic conditions. Furthermore, the human corrects many conditions of life. This is called culture, which is also an essential condition of the humankind.

I would like to transform the Playing God Metaphor to a new way of perceiving others. Genetics replaces the perspective of the Divine by the perspective of the social environment. Physicians, parents and all who know that this person is genetically manipulated compare his actual behaviour according to a socially constructed concept, which he has to fulfill and he will fulfill simply because the concept is the basic condition of his existence. This is Playing God: The inner life and the outer life fuse to one perspective. Instead of an unspecified fusion of everything by trust there is a specific fusion of our perceptions within his life by suggestion. It is a specific kind of dominance over the manipulated person.

On the other hand, we remain distant from one another, we remain divided from each other: face to face. We actually do not fuse with the other. This is exactly what I mean with confusion: We play two roles which disturb *our* self-awareness. We play God and we are persons like others. We play God and we remain persons like someone else even if he is genetically manipulated. We play the role of social interaction and claim the perspective of the transcendent. This is a confusion which disturbs *us*. And this is the reason why *we* will actually be harmed by genetic engineering.

In him we do not only see an other, but we also see our expectations in him. Thus we see ourselves in the other. This is a kind of fusion which overrides the gap of inverse kinds of evidence. Instead of a transcendent perspective in which both opposite kinds of evidence fuse, the social environment of a manipulated person represents the transcendent perspective itself. We are the other of him and we are also his inner perspective. As such we replace the transcendent perspective of his self-awareness.

3. In the face of the manipulated person we do not only perceive one person, but he is *two persons in one*. We do not meet a person who is manipulated, but we also perceive a person who is not and who will never be. Think of Kripke's theory of reference: After germ line therapy the, the patient is replaced by another one. We compare the actual person with the person who could have had lived if the germ line therapy had not happened. Without that comparison, we would not have a key of understanding what this person would be. The manipulated person

has a double face. And this double face confuses our natural relationship to him. On the one hand the manipulated person is derived from an embryo, whose plan was different. Therefore, on the other hand, the manipulated person also represents the person which could never be born because of the manipulation.

By this doubled structure we lose openness to that person because we permanently perceive two persons in one: a real person and a virtual one. But at the same time this loss of openness is accompanied by arbitrariness. Our relationship to this person is twofold; thus it becomes a question of our choice to whom we address: to the real or to the virtual person *in him*. It seems to me that moral commitments and respectful interrelationships are based on openness. If openness is lost, also moral commitments are lost.

4.5 Result

Many publications about the ethics of genetics point out that by such a technology human dominance over nature will reach the pinnacle of human history. Moreover, they stress that such technology could never be stopped or never become reversed. This chapter should have presented why this suspicion becomes apparent. Genetics change human natural conditions of self-awareness. It harms our present self-understanding because it confuses human roles of social interaction.

It is only a powerful suggestion. Actually, the natural conditions of self-awareness are still given. Human nature is also achieved in a person whose genome is manipulated. But the social environment cannot emancipate itself from that suggestion. This suggestion is essential in order to be a proponent of genetics. One must claim to know what a person should be when deciding in favour of genetic enhancement. A fundamental concept of human understanding must underlie the decision for human cloning or germ line therapy. Thus, the suspicion that genetics could never be reversed refers to social expectations and social claims about human conceptualization. If we manipulate not only a genome but also our human self-understanding, that horizon about human self-understanding could never be reversed.

This is why my phenomenological analysis about genetics is an objection against reproductive cloning and germ line therapy on principle. Even the advantages and hopes in these both possible kinds of technology could never compensate this notion.

In this chapter I have proposed a general objection against two examples of genetics. What I have not done is to object against every kind of genetics. I could imagine examples of genetics which might be quite unproblematic even if they entail ethical problems which could be solved or controlled. In principle, it is possible to compensate the ethical confusion of the so-called therapeutic cloning, the research with embryos even if they are killed. I could imagine possible ways for genetics with plants and animals which entail serious ethical problems which nevertheless might be controlled. But what will change the openness of human self-awareness should be prohibited because there are no moral reasons to change the conditions of our moral self-development.

5 Brain and Mind

Most of you will agree that we think with our brain. Our brain is responsible for our emotions, our thinking and acting. Because we have a brain we have a mind. And because we have a human brain we become responsible for our thinking and acting. Especially our moral responsibility depends on our neurological capacities.

Despite this agreement this has serious consequences you might not be ready to agree. One serious question is whether mind is a result of matter. Is it true to say that the human mind depends on a physical process? If so, then what will happen with our mind after death? Our Christian faith in resurrection or in the immortality of the soul could become relevant. Western philosophers of the modern epoch since René Descartes have not rejected that the human mind correlates to the human brain. They have also reflected the consequences of an accident with brain damage, consequences for personal capabilities and change in personal character. But despite this, modern philosophers tried to describe the correlation between mind and brain as a correlation but not an identity.

René Descartes (1596–1650) described the self-consciousness as a substance. A substance is something which needs nothing else for existing. Thus self-consciousness needs no physical matter or process for existing. Despite this, self-consciousness correlates with physical matters, especially with the brain. According to Descartes, there must be an organ as a bridge between mental and material entities. As such, the physical world has an open door for mental influence. And such the I, the human self-consciousness, could influence the human brain in order to organize human acting in life[19].

There is another thinker, Gottfried Wilhelm Leibniz (1646–1716), who describes the relation between mind and brain as entirely independent areas but with parallel structures. The mental world could never influence the physical world, but both are organized by the same structure so that a fact in the one

---

19   R. Descartes: Meditations.

world has a specific correlation in the other world. Separate but parallel structured[20].

Finally Baruch de Spinoza (1632–1677) describes the correlation between mind and brain as two modes of the same thing. According to Spinoza there is an identity between mind and brain, between soul and body. But the difference results from different perspectives which are between thinking and space. The contexts of the same thing are different, so that it seems that the same thing is two parts. Thus Spinoza tries to solve the problem about the correlation between mind and brain by making the difference between appearing and being. Actually mind and brain are the same, but they appear as if they are different. Spinoza could prove the immortality of the soul by the way: It is only an appearance when physical life ends. But the essence of mind remains valid even after death[21].

The key of modern philosophical thinking is to describe correlations between mind and brain without identifying both. Most philosophers follow this modern tradition in principle until now. We could experience and we could prove scientifically that a special structure and life history of the brain leads to a special kind of cognitive behaviour. But it is something else to say that a special neurological structure *is* a special cognitive capacity.

Neuroscience is a new biotechnological movement. In neuroscience many explorers of different sciences work together in an interdisciplinary field. Physicians, computer programmers, mechanical engineers, philosophers and theologians explore the human brain in order to describe the correlation between mind and brain more precisely. The hope is to develop therapies against different kinds of brain disease such as dementia, loss of senses (e.g. blindness), psychological distresses such as depression or schizophrenia.

One of the most interesting questions is whether there is a 1:1-correlation between brain-structures and special cognitive behaviour. Could we read what

---

20   G.W. Leibniz: The Monadology.
21   B.d. Spinoza: Ethics.

someone thinks if we perceive the actual representation of his brain state? And if so: Could we also change his thinking precisely by changing the actual brain state? Especially computer scientists are engaged in representing brain states in computers. Could we copy human capacities in computers? Could a computer "think" like a human? Could a computer "feel" like a human?

This is not only a theoretical question. It could also become a practical one. For instance, is it possible to save one's mental capacities? Suppose a very ingenious thinker who has cancer in his brain. He needs a medical treatment which could destroy his incomparable cognitive capacity. If computers could replace his cognitive skill it could be possible in principle to support the person to get his ingenious cognitive capacities back after treatment. For instance, a hard disc could compensate the neuronal defect. This leads to the assumption that mental capacities are a construction of matter. Even if a person dies, the saved intellectual capacities could work on. It is a kind of materialistic belief in the immortality of the soul.

And one important implication of that belief is the 1:1-correspondence between brain-structures and a special cognitive behaviour – which is not proven yet. It might also be possible that the same brain-state could evoke different conscious states. And vice-versa: Maybe the same thought could correspond to several different brain-states in the same person.

In spite of this, neurological research develops new and very interesting results, especially for the correspondence between brain and mind. We already know how to influence mental emotions or thoughts by stimulating special regions of the brain. Or: One could feel pain in a special part of the body, for instance in the arm, simply by stimulating a special region of the brain. One gives a weak electric impulse into the brain which is not hurtful in and of itself. But the consequence is a painful feeling in the arm although the arm is not harmed. The reverse is also true: One could stop pain or a harmful emotion by similar stimulation of the brain although the arm is actually hurt. Such, it seems, the brain constructs the awareness of the world and of oneself. And this construction – who I am and what the world is – needs not to have to do anything with the outer reality. What we think and who we are seem to depend on contingent structures of the brain.

Why is that a problem for ethics? Usually we think of ourselves as the agents of our lives, at least as moral agents for our decision-making and of our acting. But since the brain constructs our awareness of everything, we are no longer the agents of our rational judgments. We are no longer deciding what to do by rational arguments or a rational discourse. But what an open-minded capacity is depends on contingent brain structures which could be changed and influenced physically. This has consequences for what a person is. Our understanding of personality is now embedded in a materialistic frame. Personal freedom, personal autonomy appear as illusions because the brain sets the conditions for what a rational behaviour is and how a person could behave rationally.

Surely one could try to educate children to learn how to decide autonomously; how to make judgments. In this case of education, their brain will rebuild its structure. But it could become more effective to change the brain structure directly by physical means. And if so, personality appears to become a material construction. What I think, when I think I am an autonomous agent of my life, depends on brain conditions. And these brain conditions determine not only the material concept of autonomy, but also the individual concept of how I use my autonomy. Therefore, the brain is the heterogenic condition for autonomy. The point is that other humans, physicians, could manipulate the brain of a person in order to change his autonomy. And this is really heterogenic in an ethical sense.

There are some lawyers who propose not to punish criminals anymore by imprisonment. The reason is simply that criminals do not act as they want but they want as the brain determines them to do. Perhaps it would be more useful then to stimulate their brain in order to diminish their criminal energy. In some cases this has become practice already. For instance in the USA sexual criminals like rapists are treated with pharmaceuticals or even by an operation on those brain regions which control their hormones. If so, morals become replaced by medicine. One is no longer responsible for one's own actions, but whoever makes false decisions is sick and needs medical therapy. Criminal actions mutate to causal events. And morals collapse.

Since we know something specific about the correlation of brain and mind, many people use pharmaceuticals to enhance their mental capacities. According an international poll of the journal "Nature" from 2008, 20 percent of the respondents use pharmaceuticals in order to enhance their mental capacities, especially at the workplace[22]. This is called "brain doping". It seems that the consumers of such pharmaceutics do not have a special education level, social role, age, income or position in the company. It overlaps all social groups. In general, the fear of unemployment or failure are reasons for enhancing the mind capacities by drugs. In this case the heterogenics of the autonomy becomes an autonomous decision: It is the proprietary decision of oneself to become dependent on external brain stimulation.

Another point is relevant about the relation between mind and brain, at least for Christian ethics: Also religious confessions seem to be dependent on brain structures. Experiments, especially from the scientist Michael Persinger, have shown that electric stimulation at the periphery of the brain evoke mystical feelings. I cite Persinger from an article of 1983: "Religious and mystical experiences are normal consequences of spontaneous biogenic stimulation of temporal lobe structures."[23] According to new tests, people with brain damage after an accident have changed their religious and moral value system entirely[24]. Richard Dawkins states polemically that religious beliefs entail a defect of brain functioning. Religion is merely a "virus of the mind"[25].

---

22 http://www.aerzteblatt.de/v4/archiv/artikel.asp?id=65887 (discovered 11-28-09). Dtsch Arztebl 2009; 106; A 1615–8.
23 Michael A. Persinger, Religious and mystical experiences as artifacts of temporal lobe function: a general hypothesis. *Perceptual and Motor Skills* 57 (1983), 1255.
24 H.J. Markowitsch: Warum wir keinen freien Willen haben. Der sogenannte freie Wille aus Sicht der Hirnforschung; Psychologische Rundschau 55/2004, 163–168, 165.
25 http://www.cscs.umich.edu/~crshalizi/Dawkins/viruses-of-the-mind.html (Discovered 11-28-09).

Also here it is the underlying assumption that the brain constructs a virtual reality but it does not discover a "real" reality. If special brain states are the conditions for having religious impressions, the assumption suggests that there is no religious reality absent from the brain. But this assumption of construction suggests too much. The suggestion is that the correlation between brain and mind is like an identity. The premise is: "People with a specific brain damage have a specific kind of religious impressions". And the conclusion is: "The impressions of mind only refer to the brain damage but not to something else". The identity-assumption suggests that consciousness only refers to structures of the brain but not to something of the outer world. And this deconstructs consciousness.

In my opinion, this identity-assumption is too strong and only suggestive. It is not well proven because it ignores the phenomenon of consciousness. Consciousness is reduced to a representation of biological states of the brain. This is a circular argument. Because one wants to show that mental states correspond to brain states in a 1:1-correspondence, one makes mental states meaningless for themselves – and for the reference to the outer world: to what conscious states "mean". They are only representations of brain states. That is a circle. So in my opinion, we are only at the beginning of the serious question of understanding how brain and mind correlate and what this would mean for our understanding. But despite that, it seems that we must change our self-understanding and perhaps our moral foundations.

To sum up: It seems that neuroscience touches our feelings. It is like a Copernican turn which hurts human self-esteem. The original Copernican turn consisted in the remark that not the sun turns around the earth – like it seems – but vice-versa. Thus the earth and also the human beings lose their central point in the universe. Now the material claim of neuroscience intends a new kind of Copernican turn: the human mind would depart from its personal center for human self-awareness (like it seems) to the periphery of physical events. The mind seems to be only a result of physical events like all others.

In the next three chapters, I want to discuss subjects about the relation between mind and brain, which seem to be ethically relevant. First of all, I would like to discuss the claim of a materialistic neuroscience. This is rather a metaethical discussion but one which establishes the background for all further ethical claims of neuroscience. Is it possible at all to prove in principle that the mind is based on matter? And what does it mean? The second point I would like to discuss is to ask what would follow for persons with a major loss of mental capacities, like handicapped people, patients in coma or even patients after brain death? My final question is about human freedom. Do we act as we want or do we want what our brain determines us to do? Human freedom is one basic presupposition of ethics. So, if freedom is an illusion, there are no ethics anymore. It would become meaningless.

6   Could computers feel like humans? (Qualia)

So-called "qualia" are mental states which refer to how it is felt to have some mental state. Qualia refer to the subjective perspective of consciousness. The thought "This is a red dress" could be true for everyone. But how it is felt to perceive a red dress, how it is felt for me, is another matter. But it is an important one for self-awareness, especially for religious people. Christians, for instance, do not only believe in God. But their belief is accompanied by a special kind of feeling, of certitude, of silence, of happiness. I cannot know if your feelings of Christian belief are the same as mine. We could discuss whether if we feel the same, but even if we confess that we feel the same, it remains an individual feeling which could never be translated in a public language. The feeling itself, as the intensity of having beliefs, is incommensurable.

So it seems. But neuroscience has developed some proposals that other entities could feel the same, not only the same object but the same subjective intensity. One of the most impressing arguments was proposed from the philosopher David Chalmers[26]. I would like to mention his argument first: his argument why, for instance, robots could feel the same as humans if the brain could be reproduced physically.

Chalmers' argument is a philosophical one. We are not prepared to prove physically whether robots could feel the same as humans. Therefore, we need more empiric data about the brain and more technological knowledge for reproducing brain structures. So Chalmers discusses the subject by drawing out a scenario.

Chalmers sets one premise: A special mental state corresponds to a special brain-state. It is impossible for two persons who have the same brain state to have different mental states. If I perceive something as red, another person with the identical mental state could not perceive the same thing as blue.

---

26   Absent Qualia, Fading Qualia, Dancing Qualia; in: Th. Metzinger (ed). Conscious Experience. Exeter, UK 1995, 309–328.

What Chalmers now wants to show is that a mental state is independent from the biological matter of the brain. It is irrelevant whether the brain is built by neuronal bonds or whether the structure is built by silicon. A complex computer chip could replace the cellular matter of a brain, at least in principle. It is the structure alone which produces a mental state. It is not the matter. Chalmers discusses two cases in order to show that.

6.1 First Case

There is a person who is called Joe. Joe perceives a red dress. And he has a special feeling by perceiving the red dress. The redness makes a special subjective impression. Joe is compared with a robot which has the same brain structure like Joe and which is also able to give the information that there is a red dress. What many people think is that the robot could give that information without having a subjective impression of redness. On the one hand the robot gives the same information and it has the same "brain" structure as Joe. But despite this, many of us would assume that the robot has no qualia. Chalmers will show that this assumption is an implausible prejudice.

Now the experiment begins. One little part of Joe's brain becomes replaced by a silicon chip. What has not changed is the brain structure. The only replacement has to do with the matter: Instead of a natural material Joe has got a little chemical matter. After the operation we ask Joe what he perceives now in front of the red dress. What could happen now? Three cases are thinkable: Either Joe has lost a cognitive clarity. Or Joe has the same subjective impression by perceiving the red dress. Or, finally, Joe's impression of redness is a bit weaker than before. Joe perceives a red dress as before but the impression is weaker than before.

We already know from empirical studies and even from therapies that a little replacement of the brain by chips does not destroy the mental capacity at all.

Such replacements could also replace special cognitive or emotional abilities. So we could exclude the first case in which Joe's brain is entirely destroyed. Whether this happens depends on medical profession and not on principle. Thus Chalmers could concentrate on the two other cases.

Let us focus initially on the last case. Joe perceives the same red dress but his impression of redness is a bit weaker than before. If so, a paradox arises. Joe would not feel that his impression of redness is weaker. If we ask him how he would see the dress he would answer: It is red in the same intensity like before. Why does this paradox arise? The reason is that a robot whose "brain" is entirely an artifact could perceive the same like a human; a robot could give the same information about objects, as do humans. But it fails to have a subjective impression according to the ordinary assumption Chalmers wants to disprove. Therefore the first case discovers one possible implication of that ordinary assumption. One implication is that human subjective impressions become weaker step by step as long as one replaces neuronal matter by chips. The point is that there is an opposition between Joe's ability to give Joe's subjective impression, which is weaker than before, and his statement about the clarity of an intense impression of redness. Joe would not feel that his feeling is weakened. Joe's paradox refers to an opposition within the qualia.

The experiment continues. Step by step other parts of Joe's brain will be replaced by chips. After every replacement Joe will gradually lose the subjective impression of redness but he will continue to insist that he intensely perceives a red dress. Finally Joe's brain is completely replaced by chips whereas the structure is entirely saved. Now the final result would be that Joe is changed to a robot: He gives the right information about the red dress but he has no more qualia.

The suggestion of that case is that this paradox is impossible. Why should Joe not feel that he has less subjective feelings about the redness of the dress whereas he could give the right answer about the redness without feeling it? Chalmers' suggestion is: It is impossible: Either Joe feels that his impression of redness becomes weaker and weaker, but then he would not say that he intensely perceives a red dress. Or he actually remains feeling the redness, but then the replacement of a natural brain by chips would not alter anything. And then the consequence would be: A robot with an identical brain structure like a human feels the same as the human.

6.2 Second Case

The second case refers to an alternative possible implication of the ordinary assumption that robots could not feel anything. Whereas in the first case Joe loses his subjective capacities step by step, in the second case Joe's capacities remain saved after a few operations completely. But there is one later step in which Joe loses his subjective impressions immediately and completely. It is a sudden change from consciousness to a pure robotic representation without feelings. Joe would give the right information about the redness of the dress. And after the first operations he also remains to have the same intense subjective impression of redness. Here he really feels as he insists. But after the, say, eleventh operation he immediately, completely loses the subjective capacity although he insisting that he perceives a red dress. Now suppose the following case. During the eleventh operation the physicians have saved the neuronal structure of that part which they have replaced by a chip. They have also added a switch at both alternative parts: Either the neuronal part is connected to the whole brain system or the silicion chip. When the silicon chip is connected then the neuronal part is deactivated and vice-versa.

What would follow from this experiment? Physicians could switch back and forth between both options. When the neuronal part activated then Joe will have subjective impressions. When it is switched to the silicon chip Joe will not have any subjective impressions. So physicians could activate and deactivate Joe's subjective impressions back and forth. In both options Joe would insist that he sharply percieves a red dress. But only in one option Joe would really have a subjective perception of a red dress. Also here a paradox arises: Joe could never feel when which option is realized. He could never feel the switch for himself. Either he feels or he does not feel, but the switch from feeling to the absence of feeling could never be felt. And this means that Joe could never feel when his feelings are switched off.

The point is that in this case the difference between subjective capacities or the lack thereof would not be subjectively relevant. This is the result of the thought-experiment: simply because the switch between both options never becomes apparent for Joe. Joe would perceive a continuous existence as a subject although his subjective capacities are switched off sometimes. But he could not feel the change, so the subjective continuum is only an illusion. This case would make it meaningless for oneself whether he has subjective capacities or not. He could never feel the difference.

This paradox leads to a deconstruction of qualia in and of themselves. What qualia are could never be decided, because the difference between having them or not having them becomes subjectively meaningless. The conclusion is: Whoever wants to avoid this result that qualia are meaningless must concede that also the second case is absurd. And this would mean that robots with the same artificial brain structure like humans must have the same subjective feelings.

Chalmers' result implies that consciousness is entirely based on matter. Even subjective feelings, self-awareness, the subjective state of information are entirely constructed by materialistic elements. Subjectivity is nothing mysterious, it is no gift from a spirit above the world of matter. But it is simply a correlation of materialistic functioning.

6.3 Criticism

Suppose Chalmer was right. Must this result repress Christian belief? I do not think so. The fascinating point is that according to Chalmer, self-consciousness could be reproduced. My selfhood, my I-feeling could be reproduced. Therefore, one only needs to reproduce the material structures of the brain. My mind could be copied materially. And if so, why should not we think that God might create functions in that way, at the end of time? Could not we say that Chalmer is even a witness for Christian faith in resurrection? It seems to me that we could say that. It is at least a model for understanding how resurrection of body and soul could work. Perhaps it is not the best model. But Christians need not be frightened by such naturalistic explanations of human existence and human self-consciousness.

Despite this I see some gaps in Chalmers fascinating argument. The first is that we indeed have empirical support to think that human cognition is complex enough that one might not know everything about the one's own subjective mental states. A case in point is the symptomatic of the so-called "blindsight". Blindsight is a handicap in which optical information does not become a subjective impression. The blindsight person is not actually blind but she does not feel that she could see. One could show blindsight persons a sign, for instance different letters, and ask them which letter is shown. Usually they answer that they could not see anything. But encouraged to guess which letter would be in front of them they give the right answer in most cases. This is a paradox, one Chalmer has constructed in order to disprove its possibility. Now we could say that Chalmer is wrong. Thus it might rationally be thinkable that robots with the same brain structure give the same information like humans but need not feel what they represent.

But there is another point which seems more important to me: a point which shows that consciousness is more mysterious than Chalmer assumes. It seems to me that Chalmers' argument is not developed cohesively. And this is his premise that two identical brain states correspond to the same mental state. If you have the same brain state as I do – an identical structure of the brain – you would have the same mental state as I do according to Chalmers. Suppose this was true. Does not follow then that you and me are one and the same subject? How could we distinguish between ourselves since we have the identical mental state, the same qualia? Must not you think that you are me? Must not I think that I am you? that we are not different from each other?

Qualia do not only consist in feeling qualities of objects like redness. But qualia also consist in feeling oneself. If I have absolutely the same qualia as you, we both will feel identically. We would be subjectively identical then. Consequently, I would be me even if I have more than one brain. If you have the same mental state like me because you have an identical brain state I will feel myself in your body. And I will think with your brain. Our brains and bodies are fused. This would have serious consequences for our self-identity. But it seems to me that this consequence is inevitable.

To avoid such a consequence, we must exclude the I-being, the I-sensitivity from Chalmer's premise that the same brain state results in the same mental state. My individualistic I-sensitivity becomes excluded from the class of mental states. But it is not an objective content, too. So subjectivity would become something "above" mental states, that is, something mysterious. In this interpretation, Chalmers must already reject his premise in order to disprove that: that qualia are something mysterious. This is a logical mistake, a so-called circle. His materialistic argument only works if he concedes that subjectivity is a mental state. But if so, identical mental states of two persons mean that they *are* the same subject. They *are* the same consciousness. They do not only feel that they are the same subject whereas they are two subjects in fact. Subjectivity proves itself. We have no other criterion for subjectivity except subjectivity itself. So, two subjects are one and the same if and only if they feel as one and the same.

But in this case, when Chalmer concedes that subjectivity is a mental state, he could not maintain his proposition that we are dealing with *two* mental states – between Joe and a robot or between you and me. If we have the identical mental state we are not in different states. Then the differing mental state would entail a numerical difference. And we would think differently then – that is we both would have different brain states – since we think the same but do not feel the same subjectivity. We would have different brain states not only to each other but also to the state in which we would feel as identical.

In opposition to his own view, I would conclude differently than Chalmer: Under his conditions, a reproduction of my brain structure would not simply copy my mental states, but it would also change my mental states. I would now feel with two brains. My cognitive capacities would base on two brains from now on. But this does not explain anything Chalmer wants to show. The opposite is right: If my brain could be duplicated so that I have changed mental states because of that duplication, subjectivity transcends materialistic continua. Subjectivity becomes something "above" the matter. My I-feeling is not located in the brain but at best in two brains which are divided in space. Subjectivity transcends materialistic conditions. It becomes something mysterious again.

To sum up: It is one to say that cognition depends on materialistic conditions. It is something else to say that matter *is* consciousness *or explains* it. Even if there are material conditions for consciousness – and we have good empirical reasons to think that – it does not follow that we know what consciousness is then. The phenomenon of consciousness transcends its material condition even if matter is the necessary and sufficient condition for consciousness (which is not proven until now). We need not to pose a dualistic conception like Descartes or Leibniz in order to maintain that materialistic terms do not explain what consciousness is. A thought is simply something else than a brain state. Even if we read someday what one thinks by scanning his brain state we could only read his thoughts because we, ourselves, could also think. We need a mental sphere in order to compare thoughts with materialistic states. We need minds to investigate the relation between mind and brain. Therefore, the phenomenon of mind is irreducible.

I have called this chapter a metaethical discussion. It is metaethical because it has not dealt with moral norms or ethical claims. But I have discussed a premoral condition for making ethical claims. Science provokes our perceptions and even our value systems. If human the mind depends on material facts, and if the human deserves dignity because of his mind capacities, then neuroscience suggests that brain-functioning is the sufficient condition for human dignity. In this chapter I wanted to show that such a conclusion is ill-founded. It ignores the fact that human mind transcends the language of neuroscience. Therefore, the conditions for a discourse about the normative role of the brain are much more complicated than neuroscience seems to suggest.

7 Patients with serious brain damage

Since 1997 the European Unity has accepted a Bioethics-Convention to which most member-states have agreed. According to article 17 paragraph 2 of that convention it is a right of medical research to test patients with a temporary loss of consciousness even if the patients have not agreed. Such patients could become objects for medical research also if they, themselves, have no benefit. The research refers to common purposes but not to individual support of the patients. The opposite is true: Research is legitimate even if the patient gets harmed as long as the harm is only careful. I cite: "Research entails only minimal risk and minimal burden for the individual concerned."[27] Thus also shortening the life-span, if it is only at minimal risk, is legitimate. Therefore one only needs the support of a representative or of a court for such a research of patients in vegetative state.

This article of the European convention is an example of the suggestive claim of neuroscience. The suggestion is: Everything the human is and everything the human deserves depends on his or her brain. After brain damage the human right decreases. Actually one could argue in the opposite direction: One could object that a patient with a cognitive disorder needs even more protection rights than other humans because he could be harmed much more easily. But the sociopolitical climate which is influenced by neuroscience tends to the position that people with a cognitive disorder are weakened in their sensitiveness. Thus one could not harm them really.

In the last decade investigations have begun about the sensitiveness of coma-patients. What does a patient experience when he is fallen in deep unconsciousness? The results are surprising: Coma-patients are in a kind of dreamland, but not entirely. They are indirectly touched with reality. In this situation they are confronted with the actual situation in the hospital. Often they produce stress hormones during the coma or they reduce cortisol. Cortisol is an important pro-

---

27 http://conventions.coe.int/treaty/en/Treaties/Html/164.htm (Discovered 11-30-09).

tein which protects from trauma. Without Cortisol coma-patients feel heavily distressed. After wakening up many of them need a psychological therapy because they are traumatized[28].

On the other hand it is surprising that coma-patients are sensitive to impulses of the environment which lie under the line of perceptiveness. Actually one could not perceive anything. Even the conscious person could not perceive anything under that line of perceptiveness. But coma-patients could tell something about events under that line after awakening[29]. It seems to me that under these conditions medical research could harm patients much more than the European convention assumes.

Another serious problem is the case of patients whose brain-functions are irreversibly lost. This is the problem of the so-called Apallic syndrome (or persistent vegetative state). Patients with such a disease could breathe spontaneously because their brain stem is intact which organizes most of the instinctive movements of the human body like breathing. But the contact between brain stem and the cortex is disturbed. The cortex has the function of assimilating actual information from the environment. Also consciousness is located in the cortex. (This is not a philosophical statement but a medical one: The material conditions for consciousness are located in the cortex.) Because of the disturbed link between brain stem and cortex the patient could not assimilate environmental impulses anymore. Their behaviour is based on reflexes. They could feel, smell, taste, perhaps listen, but they could not organize such information in a cognitive manner.

Which kinds of rights do patients with Apallic syndrome have? I would like to discuss this question in comparison with new definitions of death. In most

---

28 G. Schelling/K. Peter: Stress, emotionales Gedächtnis undgesundheitsbezogene Lebensqualität bei Patienten nach Intensivtherapie. Neurobiologische Mechanismen und klinische Konsequenzen; in Th. Kammerer (Ed.): Traumland Intensivstation. Veränderte Bewusstseinszustände und Koma. Interdisziplinäre Expeditionen, Norderstedt 33–41, 33.

29 M. Schröter-Kunhardt: Oneiroidales Erleben Bewusstloser; in Th. Kammerer Ibid, 171–229, 189.

countries, the death of the brain is the criterion for human death. Besides, in the United States the criterion for brain death is the irreversible loss of only a part of the brain that is the cortex[30]. Even if the patient still breathes spontaneously, medical conventions pronounce him dead. Behind this evaluation, there is the assumption that only the higher "mental" abilities of the brain are the features of a human being. Thus the loss of them destroys a human. Especially the anthropology of British-American philosophy supports this thinking. The philosopher John Locke (1632–1704) pointed out that personality depends on a fluent continuum of a rational process. The consequence is that after an interruption of this process personality deflates. Present philosophers follow Locke's concept like Michael Tooley[31], Helga Kuhse and Peter Singer[32] or even the theologian Joseph Fletcher[33].

In opposition, the anthropology of continental Europe is usually based on Immanuel Kant's concept of dignity. On the one hand, the human is evaluated by his rational capacities. On the other hand it is not his actual realization of capacities. Rather the human being who *could have been rational* deserves dignity already. In this Kantian description of the human being as "animal rationabile" also the human with brain handicap, the human with brain damage and even the dead person deserves human dignity.

The difference between both anthropological concepts is based on different opinions about the relation of brain and mind. Whereas in the Anglo-American tradition the loss of brain functions is a sufficient condition for the loss of personality, in the humanistic tradition of continental Europe the membership to the human species is sufficient for dignity. The human is able to make rational judgments. Not all humans need to fulfill these abilities. But their membership

---

30   http://www.gradnet.de/papers/pomo99.papers/Junge99.htm (Discovered 11-30-09); vgl. H. Jonas: Technik, Medizin und Ethik; 236–239.
31   Michael Tooley: Abortion and Infanticide; Philosophy and Public Affairs, Vol. 2, No. 1 (Autumn, 1972), pp. 37–65.
32   H. Kuhse/P. Singer: Should All Seriously Disabled Infants Live? 159–176, 168.
33   P. Singer: Practical Ethics; Cambridge 1993, 83.

to the humankind supports their human dignity. In this concept, the brain functions are only necessary conditions for the dignity of the humankind but not for every individual. In opposition, the functioning brain is not a sufficient condition for individual personality according to the Anglo-American tradition, but only a necessary condition. Infants have a functioning brain, but their personality is weakened because their rational capacities are not still developed.

Let us return to the problem of Apallic syndrome. How should we treat patients with such a syndrome? Should we treat them as dead persons? Should we treat them as dying persons? Or should we treat them like living patients with all possible medical treatments? All three options are practiced. The first: The philosopher Hans Jonas cited cases as early as 1976, that in the USA patients with loss of cortical functioning were used as dead persons. The consequence was that they were used as organ donors. Whereas the patient was still breathing, some organs were already extracted for transplantation.

Secondly: In many Western countries, patients with Apallic Syndrome were treated according to their presumed will. Therefore, one respects their advance directive or the decision of a representative. In this case it becomes possible to treat patients with Apallic syndrome as dying people. Although they are able to breathe and swallow one could withhold nourishment until they die.

Thirdly: Actually, patients with Apallic syndrome are seriously handicapped but not dying. Their lifespan could last many years or even decades. What they need is intense care and treatment. According to the principles of physicians in Germany, patients with Apallic syndrome have the right to treatment. Their irreversible unconsciousness is not a criterion to withdraw medical treatment.

How does one decide between these alternatives? Which ethical reasons are relevant? I will focus on the subject of withdrawal and mercy killing later on. Now I will focus on the question whether we are able to say that people with Apallic syndrome have already a "spiritual" aura even if they were not able to have cognitive mental states. What I am interested in here is whether the mind and brain are as closely connected as described in the Anglo-American tradition or not. We might agree that a patient with the irreversible loss of cortical functioning has no actual mental states anymore. But in the last chapter I have proposed that mind transcends the physical language of brain states.

In the last chapter I gave reason for the proposal that my mental state changes if it is identically copied in a second brain. The same mental state in two brains leads to the conclusion that the mind is not continuously located in one brain. It is not one brain state which results in one mental state, but it is two identical brain states which result in one identical mental state. Now I would like to ask if also the opposite is true: Is it true to suggest that the mind transcends the brain even if the brain has lost its capacities for cognitive function? Does the mind of a patient with partial brain death transcend his brain state? Need the mind still be located in a brain after partial brain death?

It sounds as if that question is speculative, so I would like to turn to the question how people treat persons in vegetative state. How do they behave towards them? Which attitude would they have in face of people with serious brain damage? It seems to me that we respect the subjectivity of persons although we know that they are not able to have cognitive abilities. The "spiritual" aura of persons transcends their individualistic actual cognitive capacity. I would like to tell some experiences I have made as reverend in my parish. After the death of a person, the relatives pay great respect to the body of the deceased. They approach the deceased very carefully. They kiss him. They meet him at the final moment. And therefore, they ask for a ceremony in order to pay the dead person the corresponding respects. I remember a situation during a funeral the casket fell down into the grave abruptly. The relatives were shocked about that, although the dead person has not hurt by that fall. But it is the spiritual presence of the dead person which evokes respectful treatment, even after death.

The dead person is spiritually present because he wakes our reverence. Surely, we could behave without respect. But then, we would be ignorant of the effect that the dead have on us. Or we must decide to reject respect. Respect, however, is the first, immediate feeling we show. The spiritual aura of a dead person has an immediate authority. And we must give notice to that immediate authority even if we behave disrespectfully. Disrespect to the dead could be only a reaction to the immediate authority.

It seems to me that I need not argue in favour of this position for Christians who respect their fathers and mothers in faith with honour as Orthodox Christians do. It may be blindness in Protestant theology not to recognize the saints. At least it appears to me that we have good reasons to recognize the spiritual aura of the dead. And this aura transcends the (missing) brain states. Perhaps it is a speculative question whether the dead really have cognitive states beyond the brain states. The focus I am interested in is not a speculative one about the ontological state of subjectivity of dead persons. My focus is an ethical one. I would like to highlight the fact of respect in face of even dead persons. We would harm ourselves if we ignore this kind of respect as long as we maintain simple connections between the brain and mind. The spiritual aura transcends such easy connections. We feel obligated to respect the last will of someone. We feel obligated to treat his corpse properly, to choose the proper funeral or how we should use his belongings. Surely a connection between mind and brain remains also in this concept of respect: We know that the dead person could never interact with us as do the living. But a new space of spiritual interaction arises: We cry to the dead persons, we beg them for support. This provokes the Protestant objection against worshipping the saints: Only God should be supplicated. Otherwise the difference between worshipping the saints and praying would collapse. As a Protestant theologian, I agree in this point. But I also maintain that death and the loss of connection between mind and brain provoke a new kind of spiritual interaction.

I mention this here, because under these circumstances it harms us to treat a corpse as a simple organ donor. And it harms us when the organ donor is still breathing. Since only the cognitive part of the brain has died it harms us to use his body as a means for social purposes. And this harm is based on the spiritual aura of person which transcends their brain state.

The respect for dead persons continues after organ transplantation. We know from interviews that the recipients of an essential organ like the heart feel closely connected with the organ donor even if they did never met him and even if they did not know who he or she was. According to these interviews, the recipients tell that they feel urged to search the home of the organ donor or the place at which he has had the accident which was leading to death. Moreover they tell that their value system changes, and they think as if a second soul within their body evokes such a change. Often after transplantation the recipients report about their dreams: The dreams become more intense, they have got other qualities one could not remember about the past.

None of these examples does show anything about an ontological transcendence of the organ donor. It would be a misunderstanding to say that I would maintain a mystical opinion that the mind of the organ donor is really present within the transplanted organ. It is another point which makes these interviews relevant for our discussion: The receivers recipients the authority of a dead person. They experience the spiritual aura of the donor. They report about their respect for a person they do not know. It is an ethical fact that the organ donor is present. It might not be an ontological fact. Recipients feel *as if* the donor is present. And this is an immediate kind of respect.

My ethical conclusion is that we should search for practices which protect our respect in face of patients with brain damage, partial brain death or even death[34]. I do not object to organ transplantation in general. But I advocate practices which are respectful with the human body even if the brain functions are in disorder. It might happen that we could compensate harm against the person with brain damage through other forms of respect. It might happen that we could agree to transplantation, because we find a way how to respect the spiritual aura of the organ donor. Therefore, I would propose that an advance directive is a helpful indicator for organ donation. In my opinion, organ donation should only happen voluntarily. Conversely, an involuntary organ donation (against the will of the donor) or a non-voluntary organ donation (without any document about

---

34 This conclusion does not make the natural fallacy because it respects conditions of our morality. My remarks have already included moral implications.

the agreement of the donor) are disrespectful to the donor. They ignore the spiritual presence of the donor which transcends his actual brain states. They are based on a simple connection between brain states and mental states – a too simple one.

8 Do Humans have a free will?

Neuroscience suggests that humans do not have a free will. They are determined by the actual brain dispositions. Free will is only an illusion of consciousness, a construction for supporting and motivating human action. But the actual decision has already been made concerning what a subject would like and how she is like to act.

Such a suggestion touches ethics very hard. Human freedom is a necessary prerequisite for ethics. Only if we imagine what freedom is, could we argue ethically. And only if we are really free, could we act ethically. Thus ethics has an interest to defend free will.

The discussion about free will is slippery. It entails many misunderstandings about the phenomenon of free will. What ethics actually defends when it defends the free will seems to be something different than what neuroscience calls a free will. In this chapter I would like to give an overview of this discussion and about a plausible solution to the problem.

8.1 What is free will?

The first controversy consists of the question of the definition of what the discussion is about? What exactly is free will? The will is no phenomenon of the brain. Thus we need a concept of free will in order to argue in favour of or against free will. In neurophysical essays one finds three definitions of what free will is:

1) The free will is independent from everything[35].

---

35 D. Dörner/St. Strohschneider: Warum es keine wahren Kartoffeln gibt und auch keinen freien Willen – oder: wie man aus einem Nichts ein Etwas macht, um es dann sofort wieder in ein Nichts zu verwandeln; Psychologische Rundschau 56/2005, 236–239, 236.

2) The will is free if the agent has in mind different alternatives for acting and if he experiences his decision as freely made and if he experiences himself as the causal origin of the action[36].

3) The will is free if afterward the agent comments on his actions himself in agreement to them[37].

The first proposal seems to be implausible, although it appears to correspond to our intuitions. But let us have a deeper look at that proposal: If the free will is indeed independent from everything how could the will determine what someone wants to do? Suppose I am sitting in a restaurant and I would like to order vegetarian food. If my will is independent from everything it is also independent from my decision to eat vegetarian food. Therefore suddenly I could decide to eat meat. I cancel my first order and order meat. As long as I wait for the food my will reverses again. Now I would like to eat a vegetarian food again. Finally the service brings something vegetarian for me. But now my free will decides to leave the restaurant without eating anything.

This position is called "Indeterminism". Indeterminism describes a situation in which my decisions are entirely arbitrary. Even they are neither based on causal origins nor on practical reasons. I assume you will agree that I would be a very unfree person if I act that way - as in the restaurant scenario. Someone who is indeterminate behaves unfreely.

Thus the opposition of a determined will is not an indeterminate will, simply because in both cases the person would be unfree. The conclusion would be: We could only be free if we are determined in a special way or to a certain degree. Freedom entails a special kind of determination. Thus the first proposal collapses.

Whereas the first proposal was too strong, the second proposal seems too weak. According to this second proposal, it is sufficient for freedom that the

---

36 L. Tent: Hat er oder hat er nicht? Die Willensfreiheit des Menschen, kulturalistisch; Pschologische Rundschau 56/2005, 222–227, 223
37 H. Tetens: Willensfreiheit als erlernte Selbstkommentierung. Sieben philosophische Thesen; Psychologische Rundschau 55/2004, 178–185, 184.

person develops the subjective *illusion* of freedom: Freedom would merely be a subjective feeling with which to evaluate one's own actions. If so, others could never decide whether I am free or not. Usually you would say that I am unfree if I behaves in the restaurant scenario. But since I feel that I am free in that case, I *am* free. Your opinion becomes absolutely irrelevant.

This proposal is too weak because it ignores the question if I am "objectively" free. If freedom is only a subjective feeling about oneself as in that second proposal, it becomes apparent that the human will is free merely because we feel so. And no debate about that issue is necessary.

This leads us to the third proposal: After the brain disposes for actions the will agrees afterward to do what the brain wishes to do. The will emerges as a comment after the decision has been made by the brain. The free will as a retrospective comment combines determined aspects of several kinds. It does not combine determined aspects with indeterminate ones. The comment of the will might be based also on brain capacities; it needs not stand above the brain. It needs not be mysterious. Rather the free will combines two different kinds of determined aspects. In my opinion, this is the most persuasive explanation of what free will is. To illustrate that, I would like to draw out this position: Firstly I am going to describe an empirical experiment and secondly a philosophical argument.

8.2 The Libet-Experiment[38]

Benjamin Libet (1916–2007) was a famous researcher in neuroscience who had shown that the disposition for making decisions is taken from the brain some short moments before the will arises. His famous experiment compares the time of neuronal activities with the human self-comment about the arise of will. Dur-

---

38   B. Libet: Do We Have Free Will? Journal of Consciousness Studies 6/1999, 47–57.

ing the so-called Libet-Experiment, persons were asked to press a button down each moment they want to. In front of them there was a clock so that they could know each time at which moment they have decided to press the button. Now their brain was linked to an electroencephalogram which investigates the neuronal activity in the cortex. The researcher could compare the moment at which the neuronal activity for pressing the button had started. Conducted thus, there are three moments relevant in this test:

1) t1 is the moment at which the neuronal activity for pressing the button starts.
2) t2 is the moment at which the person is aware of deciding that she wants to press the button.
3) t3 is the moment at which the button is pressed.

The result was surprising: Between t1 and t2 there is a gap. The neuronal activity for pressing the button does not directly correspond to the will to press. Between both events there is a time distance of 350 milliseconds. Such one could say that the neuronal disposition to press the button is much earlier than the will to do so.

For many neuroscientists the Libet-Experiment shows that the human will is not free. A typical opinion is expressed by this quotation: "We do not what we want, but we want what we do"[39]. Nonetheless Benjamin Libet himself has interpreted his test differently. The point is that between t2 and t3 there is a gap again. Between the will to press the button and the action itself 200 milliseconds are gone. For Libet this time is relevant for refraining from pressing. The will has enough time to change what the person is disposed to do by the brain activity. In this interpretation, it is purposeful that the will is in the middle of neuronal activity and the concrete action. The will has the power to refrain from acting. And the will has the purpose to motivate what the person is disposed to

---

39 Critics from H. Tetens: Willensfreiheit als erlernte Selbstkommentierung. Sieben philosophische Thesen; Psychologische Rundschau 55/2004, 178–185, 180.

do. The actual purpose of the will is to give a retrospective comment for what is disposed. And this comment could even revise the disposed action[40].

This retrospective comment is important to develop a human character. It resounds to the brain in order to change or to improve dispositions. The will establishes the individual character by giving the commands about what should happen. And this information influences the brain for disposing neuronal activities in future situations.

Why this difference between determinism and Libet's own interpretation? Why does the Libet-Experiment stimulate a variety of interpretations whether the will is free or not? The reason is that the experiment leaves open what the concrete agent actually is. Is it only a subject in the present time, "now"? Or is it a person with a history, with a developed character and system of values? Certainly the persons during the experiments were people with a history, with an individual character. Such a character determines their disposition in special situations. Since we think of freedom as an indeterminate decision here and now, we also think that all alternatives (pressing or not) are of equal value. According to an indeterminate kind of view, the disposition to press or not has the same prevalence; both are preferred equally. But for people with a history, with a built character in which the interplay between brain dispositions and retrospective has occured millions times, it is not the case that all alternatives have the same prevalence.

As long as the subject of the will is a concrete person with her own character, the will has already influenced the dispositions of brain activities because of experiences in past situations. And then all alternatives are differently structured in a ranking of options. Let us reconsider the case of the restaurant scenario. Only a subject without a freely built character could not know what to order (unless it is a bad restaurant). Usually I will not order vegetarian food but rather something with chicken. It is a feature of my personal freedom that I have a preference for chicken. It is a preference which arises from my character. Without character I

---

40 B. Libet: Do We Have Free Will? Journal of Consciousness Studies 6/1999, 47–57, 51.

would be indeterminate, and if so, I would be unfree. Conversely: If I am free, I have built an individual character.

To sum up: The theory of freedom as a retrospective comment supports the theory of freedom. We should not interpret freedom as choosing between alternatives of absolute equal prevalence. Freedom entails two kinds of determination: The first is the dispositional character of brain activities, and the second is the personal character of an individual's history. Both together make human freedom of will. The free will is only valid as a personal one and never as a pure event.

8.3 Freedom means self-determination

This leads me to the philosophical argument about the free will, especially to the personal character of the free will. What is the nature of the determination of the personal character? The answer comes from the philosopher Immanuel Kant: Freedom means self-determination. I am only free if I am the author of my actions. I must determine myself in order to be free. Kant has known that the human needs not be free. There are many circumstances which make it difficult to be free. Thus the person needs to concentrate his capacities for self-determination in order to make free decisions.

I do not follow Kant in his philosophical concept of the free will. But I highlight Kant's core idea, that is: Self-determination must respect the preconditions of freedom. This kind of respect is a weak one, as is acknowledging what happens. The preconditions are not chosen freely. The free will is dependent on such conditions. So the free will has an unfree relation to its conditions. The free will could only respect these conditions. It could adopt them by a free decision. For Kant these conditions are the rules of practical reason. These reasons are purely given, independent of subjective choice. They are rationally given for everyone. But everyone needs them in order to decide freely. Thus there is a subjective dependence on rational rules. The Libet-Experiment seems to show that the free will also depends on neuronal preconditions, which come first before the will arises. Additionally, here the will could respect its natural conditions. And this respect would be a free adoption of unfree conditions.

Self-determination has to do with the free respect of the unfree conditions of the will. A free will is not absolutely free. It depends on conditions – physical and practical. But the free will is a kind of behaviour in the face of these conditions. It is a kind of interaction with these conditions. Therefore it respects them in order to influence them for a personal aim. Self-determination makes the conditions of the will meaningful for one's own decisions. It respects them in order to develop a proprietary personal character. Such development of a personal character is the second feature of a free will. The personal character is the basis for decision-making. It determines the kind of decision-making. The conclusion is: The free will depends on determinations and adopts them by self-determining them for the own personal character.

You see that in this description, free will has nothing to do with simply "doing what I want to". The reason is that such a possibility, simply to do what I want to, depends on determinations. I could only do what I want to since I have gone through the process of self-determination. And self-determination is the process of autonomously adopting the heterogenic conditions of the free will. It is the process of establishing a personal character. Unless I am a self-determined person, I could never do what I want to. But since I am a self-determined person, I will never do anything as I would like to. Rather I would only want to do what would accompany my personal character.

Now let us turn to the question if the human person is able to fulfill this kind of self-determination. Would we indeed have the capacities for self-determination? Or is this also process of self-determination determined by neuro-physical causes? Do we really have the freedom for self-determination? Or is self-determination only an illusion, and in fact we would be dominated by external natural law? In short: Is our character based on providence of the brain? Our could we partially decide which character we will establish?

Please consider what kind of answer we are looking for. We cannot prove that we are really free, because a free will is not a perceptible phenomenon. What we could really perceive is that there are living entities which have a will. But we could not perceive whether their will is freely chosen. Even if they confess to have freely chosen what they want, it is possible for them to err – natural reasons of brain dispositions could be responsible for their factual will as we have seen in the Libet-Experiment. So we could never get an empirical reason that we have a free will.

The answer we are looking for could only be a practical one, or at most a phenomenological one. We could find an answer about the structure of the will, about the structure of self-determination. Therefore, I recommend the following hypothesis: Even if we are absolutely unfree, that is, even if our process of self-determination is entirely determined by neuronal activities, our will is morally meaningful. Why? Because the will entails respect for its conditions. Respect is a moral phenomenon. It acknowledges and accepts the value or the authority in something or someone. Thus the will is always bound to morals, even if the will is not legitimate or morally wrong.

When I respect the conditions for my self-determination, I accept the pre-moral conditions for becoming a moral person. Morals themselves presuppose the free will as I have already mentioned above. Without a free will, it is not I who decides what to do. And so I would qualify as a moral agent. Conversely, since I am a moral agent because I am able to respect, I have a free will. It seems to me that this reason proves the free will.

But this is not an empirical proof. Empirically it might be controversial if humans have a free will. Also their respect could be a neuro-physical illusion similar to free will. Rather the proof is a phenomenological one. It shows the practical implications of a will. It shows the presuppositions of a will. I respect what I want to. I could only wish, like, desire, want, prefer what I respect. Otherwise I would destroy the contents of my will. I must respect what I want, at least in a weak sense (like acknowledging) that it is preferable within my will. It is not said whether it is objectively ethically preferable. But it is at least preferable within my will. Thus my will is constituted by respecting that of which my will consists.

Since I respect what I want, I become responsible for what I want. I become a moral agent because I must be able to justify what I want. Therefore it is not sufficient anymore to stress that I want to drink a glass of beer simply because of the neuronal activity in my brain. This is a categorical mistake. The neuronal activity within my brain could never justify what I want, even if this activity is the physical cause for my will. Rather, since I respect what I want, I become responsible for my will. Better: I confess to be responsible for my will. I ascribe myself as a moral agent and I become identified as a moral agent. Thus only moral reasons could justify what I want and not neuronal activities.

Here again mind transcends the brain. Even if the will is localised in the brain we need other reasons to justify our will. Even the alcoholic who wants to drink continuously is responsible for his will, although it is apparent that his will is based on a disease. Here again it is not the question whether alcohol is objectively preferable to that person but only if it is "intentionally" preferable: preferable within the will. The Swiss theologian Johannes Fischer points out that the alcoholic is responsible for his will insofar we ascribe his principle personal ability to behave according to better reasons[41]. He is responsible for abstaining even if he wants to drink. He is identified as moral agent even if he is actually incapable for acting in a correct, moral manner. Free in principle also if unfree in case – this makes the moral responsible person.

In this phenomenological sense, the free will is valid. Self-determination also functions if the brain controls human behaviour completely. Self-determination consists in respecting what is wanted. It is the self-ascription of the will and makes us responsible for our will.

---

41 J. Fischer: Leben aus dem Geist; Zur Grundlegung christlicher Ethik; Zürich 1994, 59.

8.4 Theological appendix

At the beginning of the last chapters about mind and brain, I have mentioned that theology is concerned to that relation. The question of the free will also has a theological parallel, that is the discussion between Erasmus de Rotterdam and Martin Luther. It was Luther who stressed that man has no free will in face of God.

I will not recount that dispute, but I will show how theology is involved in that discussion. Theology is interested in the line between freedom and dependence, which arises from the human himself, from his actions. Thus, theology is interested in the line between freedom and the freedom-based dependence. Freedom is self-determination, but we are not free to determinate ourselves for freedom. Remember the verse of Paul in his letter to the Galatians: "It is for freedom that Christ had set us free" (Gal. 5, 1). Christ had set us free for freedom. Freedom is not a human work but a Divine gift. On the one hand, we are capable of self-determination. On the other hand, we could not decide that our freedom is determinate. We could become unfree by our free decisions. We could become guilty. We could lose our freedom, our ability for self-determination. Thus our freedom could not guarantee to remain what it is. Freedom does not aim to freedom in itself.

The conclusion is: Neither the conditions for freedom nor the aim of freedom could be chosen. We are dependent on the conditions and the aims of freedom.

And this is the tragic character of freedom. The tragedy is that we could not refrain from using freedom, so that we must enter the situation in which we are permanently at risk of losing our freedom by using freedom. But in using freedom we behave as if we aim for freedom. Our use of freedom is not only self-determination but also a kind of self-empowerment. Since we aim for freedom, we claim to stand above the moral conditions for our self-determination. But above the moral conditions there is the position of God. We become sinners by placing ourselves in the position of God.

I will expand on what I have mentioned here. Determinism is a good example how the human will is interested in self-empowerment. The neuroscientist who highlights empirical studies, which let the human will appear as an illusion argues in a deterministic way: We do not do what we want, but we want what we do. We are programmed to do what the brain causes us to do. This is a deterministic kind of view. If so, there is no empirical reason for moral responsibility. When we are determined we are not responsible for what we do.

The result is (so-called) dialectical. Dialectical means (in short) a kind of paradox: The paradox arises by maintaining a position which immediately leads to the counter-position; and both positions are held together: Either both positions become true or none of them. This is the structure of dialectics: By maintaining a position one also supports the counter-position. Since we are not responsible for our actions anymore, everything is allowed. Everything becomes legitimate because the difference to moral wrongs collapses. Such determinism refrains from moral commitments. The result is that this is one kind of self-empowerment. The human stands above the moral because he is determined and could not behave morally.

Indeterminism is also a good example of self-empowerment. I have already mentioned that the indeterminate person is not able to act morally because there are arbitrary causes which force him to act arbitrarily. Also then the result is dialectical: Everything is allowed. Also indeterminism destroys the human capacity to act morally. Thus indeterminism supports the view that the person stands above the moral. This, too, is a kind of self-empowerment.

Both positions, determinism and indeterminism, are examples of self-empowerment. Self-empowerment is based on the aim to determine the own freedom for freedom. It wants to emancipate the moral agent from the ambivalence of freedom. Whoever stands above moral could never become unfree. Whoever stands above moral has the guarantee to remain free. Whoever stands above moral does not only determine oneself but he also determines oneself for freedom. And that is the point when moral mutates to a pseudo-divine claim. Persons who want to determine their freedom for freedom, who want to guarantee that the aim of their freedom will be freedom, claim something which could only be God's claim. They want to be as God (Gen. 3, 5). So they become sinners.

That is the special reason why theology is interested in the subject of freedom. Theology distinguishes between free will as self-determination and the self-determination *for* freedom. The second one is sin, the first one is the ambivalent structure of freedom which might lead to dependence, to guilt but also to moral purpose. The second one is a theological issue, the first one could eventually become a theological one, but at first it is a moral issue. At first the free will is the beginning of ethics.

The bio-ethical debate about the free will demands that theology highlight the difference between self-determination and self-empowerment. The dimensions of ethics and theology should be differentiated. We are not morally responsible for our freedom leading to freedom. We could not determine ouselves as free (that is the theological dimension). But we could only determine that we do freely (that is the moral dimension). But since we do so we are at risk of losing freedom. Here the moral dimension could slip to the theological dimension.

The conclusion is: Morals depend on conditions they could never establish for themselvs. One condition is the theological dimension, that God, not humans, is responsible for the final goal of their freedom. The moral agent must bear in mind that he could only remain free by Divine grace. It is not a human work but God's grace that human use of freedom leads to freedom. And it is not a human work but God's grace that even human loss of freedom leads to freedom. This is the biblical gospel of the justification of the sinner. At this point the moral dimension slips into the theological one.

But despite of that relationship between the theological and the moral dimension, it is important to distinguish both dimensions, at least for the free will concept and the concept of ethics. The human is responsible for his will because of his self-determination. Thus the human becomes a moral agent. Theology plays an important role for ethics: It holds the difference between the Divine and the human. It shows how ethics could remain what it is.

9 The problem of mercy killing

Now we turn to bio-ethical problems at the end of life. The problem of euthanasia seems to be as old as humankind in the world. Ancient documents show that in past times it was a practice to kill old or handicapped people and similarly, infants with serious birth defects. It is the typical Christian influence in ethics that this practice was banned during the ancient epoch until now.

Despite this, the discussion continues. Should we treat people medically even if they have an incurable disease? Although the ban of euthanasia has radically gained in legitimacy after the experience of the euthanasia programs of the German Nazi regime during the Second World War, the problem has gained relevance within the last three or four decades. This has two reasons.

Firstly, intensive care was radically enhanced the in the last decades. Medical research and innovations in medical practice have made it possible that the lifespan lengthens, at least for people who could enjoy expensive heath-care benefits. The consequence is that many patients survive serious diseases through new treatment methods. On the other hand, many treatments of intensive care protect the patients against death, but sometimes they could not cure them. I have already mentioned the case of Apallic syndrome (persistent vegetative state): Patients with capital loss of their brain functions could nonetheless survive for years because of the possibilities of intensive care.

The number of patients who cannot not be cured but who live longer than people in the past because of medical treatment increases. The number of patients with dementia increases exponentially simply because of the longer lifespan. Worldwide, half of the people over 90 years have dementia. It is at least 25 percent of people over 85 years[42].

In this situation, many relatives of patients hope for an early and easy death. I remember many members of my parish who thank God when a relative has died

---

42 R. Dworkin: Life's Dominion. An Argument about Abortion, Euthanasia and Individual Freedom; New York 1993, 219.

fast or even suddenly, because then he had not suffered so long. Other members of my parish disagree with the law because it is not right to shorten the suffering life of someone else. In Germany, the law tolerates suicide and also assisted suicide, but not mercy killing. The relatives of a patient with serious disease often feel guilty, because they cannot fulfill his will to die.

A majority of the inhabitants in countries of the Western world are in favour of the so-called voluntary euthanasia. Voluntary euthanasia means: If the patient decides himself that his life should be shortened, it should be allowed to kill him. Conversely, involuntary euthanasia would be killing of patients against their declared will. There is a broad consensus in the international ethical discussion that involuntary euthanasia should remain banned. Finally there is a third kind of euthanasia, non-voluntary euthanasia: killing a patient without his declared will. The difference to involuntary euthanasia consists in the following constructed assumption: The patient would agree to the killing if he was able to express the decision. But because of a handicap in communicating or even decision-making he could not do so. Examples of such possible cases are infants with multiple handicaps and serious diseases and patients with an irreversible loss of consciousness. Also in these cases, the acceptance rate for mercy killing is very high in Western countries.

There is a second reason why the problem has become more virulent in the last decades. And this reason has to do with change in the relationship between physician and patient. Many people are frightened to become a mere object of medical treatment without participating in medical decision-making. Thus the relationship has changed. In early times medical decisions are made by the physician. The relationship between physician to the patient was paternalistic. Now medical treatment is based on a contract: The physician is only permitted to treat the patient as the patient has endorsed. The paradigm of partnership replaces the paradigm of paternalistic care. Until this change, the relationship between physician and patient was based on an absolute trust in the doctor. The doctor was assumed to know better what would help and which treatment would be medically justified. Now this paradigm is partially replaced by the paradigm of partnership. It is not merely a participation of the patient, but now it is the patient himself or herself who gives the authorization for what should happen. Without that authorization, physicians cannot legitimately treat a patient. Thus the role of the patient has changed from a mere medical object to a customer.

One result of this change is the patient's authorization by so-called advance directives. An advance directive is a written document which prescribes the patient's will for times he could not actually express it because of unconsciousness. Many advance directives order a withdrawal of treatments when the intensive care would not have a benefit except pure survivial whereas life becomes suffering. I will discuss the problem of advance directives later on. Here I only mention that the new relationship between physician and patient supports a new discussion about the legitimacy of shortening the patient's life.

Shortening life and shortening life are not the same. There are three different instances of shortening a patient's life. The terms are passive euthanasia, active euthanasia, indirect euthanasia. What do these terms mean? One could shorten a patient's life by withholding essential treatment. In this case, the patient will die because physicians do not further treat the sickness. Here it is the sickness which causes the death primarily. Nonetheless, the withholding also causes death, at least secondarily. Thus, the question remains whether physicians are morally responsible for the patient's death because they have withdrawn possible essential treatments. This kind of passive euthanasia gains more and more acceptance in justice and medical ethics.

Active euthanasia is the direct intervention in the patient's life to kill him. The patient would not die or he would not die as quickly if that intervention was omitted. Here there is a significant causal and moral responsibility of the doctor for the patient's death. Active euthanasia is the most controversial kind in the ethical debate of shortening the patient's life.

The third kind, indirect euthanasia, is based on the Roman-catholic conception of double effect. According to Thomas Aquina it is legitimate to fulfill an action for good causes even if it results in an effect which is not desirable. In medical ethics, this takes place, for instance, when a patient needs pharmaceuticals in order not to suffer. The dose of pharmaceuticals might be necessarily so high that the risk is given that the patient will die. Despite the risk, the physician is allowed to treat the patient by that high dose of pharmaceuticals because he does not want to kill the patient but only to help him not to suffer. This is the double effect: On the one hand, the physician aims to benefit the patient – that is not to suffer – on the other hand, a second effect arises, that is death, which is not desired but causally connected with the aimed benefit.

In discussion there are many authors who are skeptical whether this difference between the three kinds of euthanasia is morally relevant. Especially consequentialistic thinkers object that in all three cases there is no difference in causal responsibility for the patient's death. Consequentialism is an ethical concept which evaluates actions for the best consequences independently from the moral character of the agent. According to the consequentialistic kind of view, an action is good not because one follows good moral principles. Rather an action is good only because one has the best intentions.

Thus, according to that conception, the difference between active, passive and indirect euthanasia collapses. In all three cases, the physician's behaviour is a causally sufficient reason for the patient's death. And in all three cases the doctor knows what will happen through his behaviour. Consequentialists conclude that one kind of euthanasia legitimizes all other kinds too. The most famous consequentialist ethicist is Peter Singer. His most famous published work is his book "Practical Ethics".

In opposition to a consequentialistic position there is a deontological kind of view. Deontologists argue that not the effect makes an action morally significant but the attitude with which the action is executed. The theologian Gilbert Meilaender[43] is a follower of a deontological conception like most other theologians. There is indeed a difference between these three cases in virtue to the attitude. Meilaender gives an analogous example: "Just as the soldier going on suicidal mission does not choose to die but, rather, to live in a certain way, rec-

---

43 G. Meilaender: Bioethics. A Primer for Christians; Michigan 2005, 70.

ognizing that to live in this way may mean not to live as long". Meilaender concludes: "To reject or withdraw treatment because of its burdens is still a refusal of treatment, not of life"[44].

A consequentialist could agree that the attitude is different in these three cases of active, passive or indirect euthanasia. What he rejects is that the attitude is morally relevant. Why should that be? Why should the attitude make an action significant as long as in effect and in knowledge about the consequences all three cases are similar? Conversely, a deontologist could argue in the opposite direction: Why should the effect be the sufficient condition for making a moral significance of an action? Is not the attitude the kind of respect which makes morals exist?

Are there moral reasons to decide between both of these concepts? Could we ethically decide whether a consequentialistic or a deontological conception should be morally preferred? Many ethicists, from John Rawls to Peter Singer, have already tried to do that, but it seems to me that they set preconditions in their arguments which the counterpart need not accept. Thus it seems that no position could be held without prejudice. We must concede that both criteria – attitude and effect – are important features of ethics. Moral behaviour which would never achieve a good effect would be nonsense. And moral behaviour which would ignore the attitude of the agent to someone would ignore an essential resource for moral respect. Such a moral effect could not be respected. And also this sounds absurd. The question could only be which criterion plays the more important role if both criteria come in conflict with one another.

But it is another matter whether such a conflict is valid in the three cases of euthanasia. Is it right to say that both criteria compete? Or should not we say that both criteria are coherent to each other? It seems to me that there need not be an actual conflict. It might be that according to the effect, all three cases are similar. But then the moral significance of each of these three cases could not be decided simply by the criterion of effect. As long as they are accepted as similar

---

44   Ibid.

in order to give reason why active euthanasia should be legalized, one could not prefer active euthanasia. One could not reason why active euthanasia is banned in medical treatment. No one would miss active euthanasia if passive euthanasia has already a similar effect which should also give reason why active euthanasia should be permitted.

Consequently, one needs the second criterion, the difference in attitude, for making the difference significant. Since we argue that both criteria are important for ethical decision-making, the conflict between deontology and consequentialism breaks down.

A consequentialist could object that the three cases might be similar in effect but they are not of equal effect. It might be that killing is more effective than withdrawing a treatment. If so, an inverse ranking could arise between these three cases according to deontology and consequentialism. Whereas the attitude of killing makes active euthanasia immoral according to the deontological kind of view, killing is more effective, frees from suffering immediately and more quickly than passive euthanasia. In short: The consequences of active euthanasia would be better.

The point is that we have to explain why the attitude is significantly different between killing and letting die. Why is the attitude different between killing and giving an injection against suffering which would indirectly lead to death? Perhaps, and this is my suspicion, the opposition between both ethical concepts only arise because the term "attitude" is not well founded. To which object do we address moral attitudes? Is it human life or is it the human herself?

Since the moral attitude refers to human life, it is true that in all three cases human life is harmed by the physician's behaviour. Therefore one could not distinguish between the three cases of euthanasia. One could not argue that in the case of active euthanasia, there is no respect for life, whereas in the other two cases there would be. In all cases, the physician could act without respect for life. And in all cases he could act in respect for life, knowing that life is not eternal, and sometimes it might be morally justified to help the person die. To shorten life does not automatically mean that there is no respect for life.

Such it might seem that according to the deontological perspective the attitude refers to the human, to the person concerned. A deontologist could argue that the person is dignified by a moral attitude. And this attitude could also lend dignity to a person if she is dying. Respect for the person includes respect for her dying. The withdrawal of treatment could be in accordance to her dignity because it respects her natural process of dying which is embedded in the dignity of the person. – I agree to that point, but it seems to me that this argument fails to distinguish between passive and active euthanasia. Additionally, in active euthanasia, it might be possible to intervene, shortening life in spite of a moral attitude for giving the person dignity. In short, one could kill in love. One could kill out of respect for the dying person. And one could kill out of respect for the patient's desire to die. There might be no significant difference in attitude between killing and letting die.

Indirect euthanasia supports this assumption: Since it seems legitimated by the deontological perspective to kill someone because of the double effect, one actually intends to save the patient from suffering and one concedes that this is only possible by killing. The attitude for saving not the patient's life but his dignity legitimizes indirect euthanasia. But if so, then active euthanasia could also be right by the same criterion. It is the *identical* criterion to save one's dignity, to save the patient from suffering, by conceding that this leads to death. I conclude that also a moral attitude for the personality of a patient is not sufficient for making significant moral differences between these three cases.

For me it is a theological reason which makes the difference between killing and letting die significant. According to the 10 Commandments "Thou shall not kill!", killing is against the Divine Law. However, in a pluralistic world with different religions and lifestyles, this theological argument in this kind of view does not seem persuasive. There is also a reason of theological ethics why the pure reference to the Divine law is not persuasive: As I have already mentioned in the first lecture, it is not ethically sufficient to refer to the Divine Commandments for reasoning ethical claims. Ethics is a kind of reasoning. To say that something is morally prohibited, simply because it opposes a law, provokes the question whether the law is morally justified. And therefore it is not sufficient to answer that the law is justified because it is a Divine law.

1) One could be skeptical as to whether it is – not only atheists could wonder whether the Commandments of the bible are Divine since historical investigation of the bible reveals different sources for biblical claims, even from other religions.
2) But even if we accept that the law is a Divine we need not conclude that it is eternally right. St. Paul has argued in his letter to the Romans and the Galatians that the biblical law is no longer relevant for Christians.
3) But even if we accept that the Commandment never to kill is relevant for Christians, we need ethical reasons for justifying that claim. Therefore, it is not sufficient to recount that it is a Divine law. Rather we must find the ethical substance for that claim. What is the ethical substance for the ban of killing even in the case of euthanasia when one accompanies a person with fulfilling her will to kill her?

It seems to me that this reason is not discovered until now. We have not yet the reason why killing with the same attitude as in letting die is generally wrong whereas letting die could be justified. And we have not yet found the reason why the attitude in killing is different from the attitude in letting die. The crucial point is: Could we respect God in killing a person dignified by our moral attitude? What will change in our respect to God if we allow active euthanasia?

In my opinion, the problem of euthanasia puzzles our moral feelings. Here again, like in the chapters about genetic enhancement, our moral feelings are touched by the responsibility to decide for or against human life. In the following two chapters, I will show which feelings are involved and which moral consequences should be drawn. For me, the issue starts earlier: It does not start with the question whether euthanasia respects human life. Rather, the problem already arises how we respect life when it begins to end. Withdrawing a treatment entails respect for the process of dying. But here I cannot see a significant difference to active euthanasia, at least: it needs not be a significant difference. I do not discuss the possible misuses of killing persons in disrespect. But I also do not discuss the possible misuses of withdrawals of treatments in disrespect. Misuse might happen in all cases. But this is not a deontological question how to behave in the face of such possible misuses. It is simply a consequentialistic question[45].

Nonetheless, the problem of euthanasia puzzles our moral feelings. In my opinion, this is the actual moral challenge facing modern medicine. Indeed modern medicine has come to a point in which decisions about the end of life become more and more relevant because of the tragic cases of intensive care. However, this uncertainty does not only refer to active euthanasia, but to the general issue of end of life decisions. We have become responsible for the end of life of our neighbour. This is puzzling because it contradicts our moral intuition that life is not at part of our moral decision-making. Actually, we confess that life is a gift of God. Now the question becomes crucial: Could we fulfill our responsibility for end of life decisions without touching the Divine authority who alone is the Giver and Savior of life?

In the next two chapters I will discuss two crucial points: First, the problem of attitude in respect to end of life decisions refers to the relationship between life and eternity. If we respect someone as a person we do so independently from his actual life-situation. Respect to a human does not end after he has died, as I have already mentioned two chapters previously. But that does not mean that a person's life-situation has nothing to do with our respect for him. Thus the first point refers to the question: How could we respect a person in our "physical" conditions of living together even if that person's life is at its end? This is the first point I would like to discuss in the next chapter.

The second point: the confusing feelings refer to a double meaning of guilt. Here again I reaffirm my statements about the difference of guilt and sin in the chapter ago. But I will reformulate slightly. Without a doubt, it is a problem of

---

45 Besides, also the so-called "slippery slope-arguments" entail a consequentialistic perspective. If we enter a slippery slope , it is not merely based on moral terms but also on empirical and historical circumstances. (J.A. Burges: The Great Slippery-Slope Argument; Journal for Medical.Ethics 19/1993, 169–174).

ethics how we could avoid guilt by killing or withdrawing essential treatments. This is an ethical challenge. But it is no longer a problem of ethics why our moral feelings are puzzled even if we have good reasons for justifying shortening human life. Even if we have good reasons for fulfilling responsibility for the nature of a patient's death, we may feel guilty because of our decisions. We may feel guilty because of the tragic enormity of end of life decisions. We may feel guilty, because we are acting out of a dilemma. We are responsible for that dilemma because intensive care is a result of human culture. But we could not refrain from intensive care because otherwise we would destroy the benefits of that kind of medical treatment. Thus we cannot choose not to act within that dilemma.

Note that this term "We cannot choose not to act within that dilemma" is similar to the description of sin by Augustine (354–430). The term "We are not able not to sin" describes Augustine's concept of original sin. My view is that we enter the theological subject of sin (and even grace) when we decide about one's life at its very end. But the subject of sin and grace is not at first an ethical issue but a dogmatic problem or an issue of Christian pastoral care.

In the next two chapters I will expand upon these two interpretations about the moral attitude of euthanasia in face of God. I will begin with the relationship of life and eternity and then turn to the difference of moral guilt and sin.

10 Eternal Dignity

There is a reluctance to kill a person which has something to do with the religious respect to God. And this moral attitude refers to one's dignity. If we respect one's dignity we feel compelled not to harm that person. The urge arises from the feeling of one's dignity. This urge provokes our attitude of respecting his dignity. It is not the same but at least similar to the religious attitude of worshipping God. The churches insist that human dignity is based on the biblical conception that the human is the image of God. It seems to me that the biblical concept of the image of God has some other implications. But the insistence of the churches highlights the intimate relationship between the religious attitude to God and the moral attitude to one's dignity.

In my opinion, dignity shares a character we ascribe to God, that is eternity. Therefore I would like to discuss a case in point.

10.1 A metaphysical presupposition of Dignity[46]

Given a suffering patient whose disease cannot be treated anymore. Medicine will not have a benefit for him but it prolongs his burden. This is a situation which often causes the following statements:
1) a. „The life of this patient is without dignity". Or b.: „The patient has been stripped of his dignity". Or c.: „The patient has lost his dignity".
2) „*Consequently*: Let him die". Or: „*Consequently*: Let us kill him".

The three sentences of No.1 do not express the same. But in connection with No. 2 they suppose the same: They suppose that the patient will not lose his dignity *after dying*. So: The patient will exist after dying. The reason for this is that kill-

---

46 L. Ohly: The Metaphysics of Mercy Killing; in: D.P. Georgi (ed.): Bioethics. Need for a Paradigm Shift; Amchit 2012, 135–143.

ing or letting die is presumed to give his dignity back to him. No serious position uses No. 2 in order to stress: „For, when the patient has lost his dignity, we can do everything we want to do with him. We cannot harm him anymore, so we can do everything with him according to our whim. So let him die or better: Let us kill him". This inhuman statement is not meant. Rather No. 2 is proposed because assisting him in dying would be in accordance with his dignity. But it can only be in accordance with his dignity if dignity does not cease with the death of that patient. Otherwise one would think in the same way as the inhuman statement above: „Let us kill him in order to have no more duties to his dignity". This sounds absurd and contradicts No.1, because according to No. 1 dignity is already lost. So, the two sentences suppose that on the one hand the patient has lost his dignity *and on the other hand he has not*, because the loss of his dignity enforces us to assist him dying in order to bring his dignity back to him.

But how can one have lost his dignity and nonetheless keep it? Is that not a contradiction? It is not, if the term „dignity" has a double meaning. There is a connection between both meanings of dignity but they are different. In the first sentence, the term supposes a "mundane" fact about something which can be hurt or entirely eliminated on earth, during one's life. The second sentence supposes that dignity is protected and cannot be eliminated eternally. For it also commands respect if dignity in its first meaning is lost. The concept of dignity entails a special connection between two terms of dignity. On the one hand dignity has a "mundane" meaning, on the other a transcendent one. Thus if someone argues by dignity she implies physical and metaphysical presuppositions. Metaphysics means in this context: One refers somehow to a second reality which transcends the obvious reality of our objective world. In the physical world, someone has lost his dignity, but in a transcendent reality beyond the physical world he keeps his dignity eternally. Thus he remains the right to get his dignity back also in the physical world.

One could object that the difference between these two terms of dignity only consists in the moral difference between „be" and „ought to be": The first one describes a human state whereas the second one prescribes an ethical challenge.

For instance, the patient has actually lost dignity but he ought to get dignity. – But why *should* we assist someone to get the dignity back which he has lost? If he has lost dignity he would also have lost the claim to get dignity back. Here the difference between "be" and "ought to" could not explain the logic of these two sentences[47].

Let us consider the following two differences:

1) If I have lost something of my own because I was not careful enough (for instance, my key of the guest house), then no one has the duty to give it back to me or to help me. It would be nice if someone helps me search, but no one is obligated to do so. I have no right that others should help me search. In this case, there is no connection between „be" and „ought to be": It is not useful to say that I would have the „right" to get it back because if I had such a right, someone else would have the duty to give it back, although he is not responsible for the loss.

2) If a thief has stolen my key, I indeed have the right to get it back, because it is mine in spite of the loss. Now suppose the thief has not only stolen but also destroyed my key. Then my right for possession would even preclude the non-existence of the key. I keep the right for possession of my key even if it is destroyed. I keep the right to demand him compensating my loss. In this case the terms "be" and "ought to be" can be differentiated. Something of my property has been destroyed but I keep the right of possession.

Now let us compare this to dignity: If someone had destroyed my dignity, I could not claim any rights to get it back or to be aided. The reason is that dignity is not the same kind of thing as keys which could be my property. Dignity is necessarily connected with claims, so that if dignity does not exist no claim for dignity will exist. And if there is a claim for being respected with respect to one's dignity, one has it in fact. The difference between "be" and "ought to be"

---

47 Furthermore: In context of euthanasia the metaphysical presupposition entails the assumption that only death would bring the patient's dignity back. Because his dignity *is* lost he *ought to* be dead. Here the reference to a second reality beyond the physical world of life is obvious.

cannot be divided into two parts of respect to dignity, because the right to have dignity is immediately connected with dignity itself. But despite this, the difference is not useless because if my dignity is destroyed by others, they are responsible for the loss. They have harmed my rights and they (or others) have the duty to compensate my loss. So there is a connection between "be" and "ought to be" but we cannot divide them into two separate parts. The consequence is that the "ought to" is a part *within* the „being" of dignity. And if someone loses his dignity but he keeps the right to get it back then this right must be a part within a *second* „being" of dignity which transcends the "mundane" character of a lost dignity. This transcendent dignity cannot be lost[48].

10.2 A second metaphysical presupposition of dignity

There is a second metaphysical implication implied by these two sentences. It is noteworthy that according to these sentences, we respect one's dignity by shortening one's life. Such a suggestion must presuppose that the dead person also still has dignity. The crucial point is that according to these two sentences, human dignity prolongs one's death. The consequence is that we remain obligated to respect his dignity and to treat him with dignity after dying. Furthermore, the two sentences above presuppose that shortening one's life restores one's dignity. Although he is not living anymore and he is no natural being anymore, he re-

---

48 Let us consider this hypothesis from a contractual perspective: One could argue that dignity depends on a social agreement. In my opinion, this might only be the ultimate condition of *discovery* of human dignity but under no condition dignity itself. If agreement were a condition of dignity itself and someone entirely lost his dignity, then this loss would depend on the society's agreement about his loss. How can the society agree that he „should" get his dignity back without agreeing that then he still *has* dignity in another point of view? Thus: Also if social agreement is the basis for dignity, two terms of dignity will be drawn out. And one of them would be presupposed as metaphysical by this agreement.

mains a human being who deserves human dignity. Thus, it seems, he would be a supra-natural being who deserves dignity. The presupposition entails the thought that only his dead existence corresponds to his eternal dignity. His death becomes a kind of existence. Although after death he is non-existent under natural conditions, he remains existent under supra-natural conditions. On the one hand he is non-existent under physical conditions, but on the other hand, he remains existent under metaphysical conditions.

These metaphysical conditions offer the ethical framework for dignifying him by shortening his life. His death corresponds to his dignity because by dying, he leaves the mundane area which harms his dignity. By his death he arrives in a different area in which his dignity becomes secured. Whereas his natural dignity threatens his dignity or even destroys his dignity, it is saved under metaphysical conditions. The ethical conclusion is the following: As long as we are obliged to respect his dignity, we should support him by turning to the metaphysical area in order to save his dignity, which is destroyed under physical conditions.

The metaphysical implication of the argument in these two sentences above is apparent. Here again the term "dignity" possesses a double meaning: physical and metaphysical implications. And here again we cannot reinterpret this double meaning by the difference of "be" and "ought to". Why should we shorten one's life in favour of his dignity? The only reason would be that he deserves dignity. But his dignity is saved only under the condition that his existence "survives" his death. Otherwise it would never remain "his" dignity. It would never belong to "him". Therefore it would not be an ethical reason to respect "his" dignity by shortening his life. The conclusion is: Shortening life is only ethically justified if we ascribe metaphysical existence to a person who deserves dignity also after death.

Connecting both metaphysical implications leads to three different interpretations of human dignity:

1) a mundane state which might be lost under special circumstances;

2) an eternal gift which refers to all situations of the human, especially to those in which the mundane state is mistreated or even destroyed; here the term "eternal" means "timeless";
3) an eternal state one refers to when arguing that only death will bring dignity back. Here the term "eternal" means "in the future" – after death.

10.3 Non-metaphysical approach to end of life-decisions

These kinds of metaphysical implications within the concept of dignity are similar to religious hopes for immortality of the soul or in the resurrection of man. Similarly, dignity transcends the real situation of the mundane world. It seems that the concept of dignity is based on the same understanding of human existence similiar to religious beliefs about a second reality beyond the world. However, in a pluralistic situation, concepts of second realities beyond the world become implausible. This makes the concept of dignity implausible, too. There are several ethicists who reject the claim of human dignity because of its metaphysical implications.

However, in this section I will show that such metaphysical implications become inevitable even if one tries to avoid the claim of dignity. Especially in the case where end of life decisions metaphysical implications become unavoidable.

Which are the strategies for translating the meaning of dignity into a non-metaphysical context? How can metaphysical implications be avoided at the very end of life? Several alternatives are possible.

1) One could reject the whole concept of dignity. The most popular example for this strategy is utilitarianism[49].

---

[49] J. Harris: The Value of Life; London 1985; P. Singer; Ibid.; H. Kuhse: The „Sanctity-of-Life Doctrine"; in: H. Kuhse/P. Singer: Individuals, Humans, Persons. Questions of Life and Death (ed. E. Morscher, O. Neumaier); St. Augustin 1994, S.33–65.

2) The other strategy accepts the double meaning of dignity but wants to illustrate the transcendent part in a non-metaphysical way. Cases in point are concepts linked to Kantian philosophy which describe dignity as an ideal which has no real ontological ground but only a moral and political *function* approximate to this ideal. The German "Diskursethik" from Karl Otto Apel[50] and Jürgen Habermas[51] leans in that direction[52].

I am skeptical whether these concepts will be successful as long as one of their most respected elements for decision-making is the human will. One can show that the human will also implies metaphysical suppositions.

According to the non-metaphysical concepts above, mercy killing is morally valuable if someone wants to die (e.g. utilitarianism). Or they show that no patient is able to wish his death because he has higher will to live which is only morally relevant (e.g. Kantianism). But also the latter does not deny that human will is fundamental for moral decision-making. Perhaps every moral concept is based on human will, affects, interests and motivations to do that which is good and right. But since human will is the basis for moral values and norms, these concepts get a metaphysical boost. In his writing "Unscientific Postscript", the Danish philosopher Sören Kierkegaard (1813–1855) has stated that every human interest can be translated in the only interest in oneself. And this interest is an *eternal* one[53]. But he failed to give an exact proof for his statement, because his philosophical concept excludes logical proofs as aesthetic and non-ethical[54]. To prove his statement, one must be aware of his own existence and only then it would follow, not logically but „existentially".

---

50 K.-O. Apel: Transformation der Philosophie. Vol. 2: Das Apriori der Kommunikationsgemeinschaft; Frankfurt a.M. 1976, S.358–435.
51 J. Habermas: Erläuterungen zur Diskursethik; Frankfurt a.M. 1991; J. Habermas: Moralbewußtsein und kommunikatives Handeln; Frankfurt a.M. 1992⁵.
52 D. Böhler: Menschenwürde und Menschentötung. Über Diskursethik und utilitaristische Ethik; Zeitschrift für evangelische Ethik 35/1991, S.166–186; A. Cortina: Diskursethik und Menschenrechte; Archiv für Rechts- und Sozialphilosophie, 76/1990, S.37–49.
53 S. Kierkegaard: Concluding Unscientific Postscript, Princeton 1944, 25.
54 S. Kierkegaard: Concluding Unscientific Postscript, Ibid., 99.

I would like to give a logical proof because one needs not share Kierkegaard's philosophical framework in order to discover implications of eternity in every interest. Let us consider the sentence „I would like to die". This sentence is not the only one which can be illustrated as metaphysical, because only the logical form of human will decides. But this sentence is the best illustrator.

If one wants to die, he wants to be satisfied by an action or event, because every interest searches for satisfaction. And if the result is fulfilled, the interest simply ends. Despite the end of the interest, another interest continues which will not end: This is the subject's interest in oneself. The subject has an interest in oneself because the subject wants to be satisfied by the result of the preferred event. So the subject presupposes to exist afterward also when the interest in something comes to an end. The subject wants to override that interest. Now in the sentence above, the subject wishes to die. However, also in this interest the subject must wish to exist afterward: He (or she) presupposes that she *survives* that interest because she wants to be satisfied by that interest.

Utilitarians support this proposal against their non-metaphysical confession – unfortunately against their non-metaphysical concern. Nonetheless utilitarians stress that for some patients death is „better". But if so: What could the term "better" mean, if beyond this life nothing exists which could be better? If nothing exists beyond the physical life, no one could compare a state beyond physical life with life which seems to be worse than dying? – My answer is: The concept of interests entails the metaphysical idea of the interest in oneself. And this interest in oneself overrides all other interests and all other states contained in all other interests. The interest in oneself even overrides life and death. Whoever wants to die, has an interest in himself. And according to that interest his subjectivity, his I-feeling, overrides life and death. Only then could death appear to be better than life.

I conclude that an ethics which aims to satisfy interests must bear in mind the metaphysical concept of the interest in oneself. Every interest entails the interest in oneself. Therefore it transcends the satisfaction of interests. It overrides all other interests. It transcends to a metaphysical world in which everything is different except the I, the subject himself.

My proof about the metaphysics of interests is still not sufficient. One could maintain the following objections:

1) From a presupposition about eternity, real eternity may not follow. Even if every interest implies an interest in oneself and even if the interest in oneself is perceived as eternal, it does not follow that the subject will be eternal.
2) Not every interest begs fulfillment.
3) Suicidial interests are possible in that way: „I (now) want not to exist *afterward*". Although every interest might have an interest in oneself in the same time one has the interest, it does not follow that someone could not have an interest not to exist afterward.

I accept the first objection and reject the others. However, even if the first objection is right, it does not solve the dilemma of non-metaphysical ethics. If human will is respected, is the main ethical purpose. Then we cannot ignore the ontological possibility that subjects exist after their life unless we show that such a possibility does not exist. I think such a proof cannot be made unless one shows that a dead person could not have mental states anymore. Therefore one might identify mental states with brain states. Dead persons have no brain states. Such they could not have mental states. I have mentioned in my lectures about brain and mind that this is not a proof against a metaphysical existence of subjectivity. The mind transcends the brain. Thus we could not identify mental states with brain states[55].

The conclusion is: As long as we are not able to exclude ontological possibilities for eternal subjects, we could only exclude our respect for such eternal sub-

---

55 Several chapters above I have already mentioned that this is not only an epistemic problem but also a logical one, because the human discovery of an adequate 1:1-correspondence between selfhood to bio-physical facts must include the discovery itself. But then the I must discover that his discovery really corresponds to his thinking about this correspondence. And this brings the subject to an infinite regress so that the I cannot prove what he has stated.

jects or our respect for the interest in oneself. I wonder whether an ethical concept remains non-metaphysical, even if it rejects respect for metaphysical possibilities. Such metaphysical entities similar to eternal subjects remain ontologically possible, so that we would hold a prejudice disrespecting what remains possible and what is necessarily stated in every interest: the interest in an eternal self. This prejudice not to respect what is nevertheless possible is a judgment about metaphysical possibilities. On the one hand, it is a non-metaphysical prejudice, on the other it deals with metaphysical possibilities as if only the physical reality is possible. So non-metaphysical thinking changes to a kind of metaphysical replacement.

Moreover, the ethical question arises: Are we permitted to disrespect the interest in oneself especially if it accompanies every other interest? By which moral reason could we disrespect that which is morally presupposed – an eternal subject – and what is ontologically possible?

My opinion is that *there is no good moral reason for rejecting the eternal interest of someone in oneself*, since respect for human will is the main, absolute purpose of non-metaphysical ethics. Rejecting respect for the eternal selfhood of someone, which is a necessary condition for having interests, is only acceptable if it is juxtaposed against a higher ethical value. But I do not think that there is one higher reason to reject respect for eternal interests unless the non-metaphysical strategy itself was such a higher reason[56].

Let us address the other objections. I agree partially with the second, that there is one interest which cannot be satisfied but only be respected. This is the interest of a subject in herself, and that is the reason why it is presupposed as an eternal one. The subject cannot want to be satisfied in her interest in herself. So she wants to be eternal. The interest in oneself does not know the point of satisfaction. The second objection has to show that there are interests that do not entail the will of satisfaction but nevertheless should end as an instance of the

---

56 Then it would also become metaphysical because it would be a final reason for ethics at all.

interests themselves. Therefore the third objection could be such a remarkable interest. So let us turn to the phenomenon of suicidal interest.

In my opinion, it cannot be proved that suicidal interests do not entail the interest in oneself. Given that the subject thinks about herself in a way which does not prefer her future, she only thinks of herself as a present entity ignoring the question about her future. Then she cannot want that *she* will not exist *afterward* because according to the condition she understands herself as an entity without thinking of any „afterward".

Since the objection accepts that every interest implies the present interest in her own subject then the subject cannot want her end, because she cannot think any end of herself – simply because she thinks herself only in presence. In this case the sentence „I do not want to exist afterward" is a contradiction according to the premises of the objection. Or the sentence entails two concepts of „me". One of them is the process of having the interest in herself. This interest could never entail the wish for an end simply because the subject cannot want to be satisfied about herself; nor could she think of any time later because she thinks only in the present time-dimension.

The second concept means a person frankly can come to an end, who can die and who can want to die. In this second meaning, one could also want to destroy oneself. But even then, the interest in oneself will not be reduced in a non-metaphysical kind of view. For there will be one moment at which the present interest in oneself replaces the interest in destroying oneself. Therefore a suicidal person can want to die, but he cannot want to eliminate *himself*. He cannot want to destroy his selfhood because he needs his selfhood for his interest for dying.

"I (a present state) have the interest to destroy myself (as a future being)."

Suppose the contrary: Suppose a subject wants to die and also to eliminate oneself. Here the meaning of selfhood is twofold. As long as I have the interest of dying I must have an interest in myself (as a present state) because otherwise I could not have the interest of dying any longer. As long as I hold this interest in myself (as a present state) I prolong the interest in myself in an undetermined future. I have the interest in myself until I have reached my goal to destroy myself.

But there is a point in which both meanings of selfhood meet. That is the moment of suicide: "I (a present state) have the interest to destroy myself (as a *present* being)". And at this moment a contradiction arises. Either I have an interest in myself; but then I could not want to destroy myself. Or I destroy myself, but then I could not intend to destroy myself. In a non-metaphysical kind of view, suicide could only be an unintended event. Frankly this sounds absurd.

To avoid the contradiction, I must concede that destroying myself as a *being* has nothing to do with myself as a *state*. So, my interest in destroying myself as a being will not touch my interest in myself as a present state - my interest in myself will also prolong the intended death of myself because of the metaphysical implication of having an interest in oneself. In this case, we have to distinguish between the desire to end one's own life (as a person) and to end one's own subjectivity, which is impossible to think.

To sum up thus: If ethics is based on interests which should be fulfilled, then only intellectual ignorance will protect a system of „non-metaphysical ethics". That means: Non-metaphysical ethics is a contradiction. Or it only functions by being *silent* about ethical problems at the end of life and about selfhood, self-consciousness and self-confidence. But frankly: Such a specialized system would be mostly useless – especially in medical ethics.

At the beginning of this section I have stressed that the concept of dignity entails a transcendent element. As I have pointed out here in discussion to non-metaphysical concepts of ethics, this transcendent perspective is inevitable since one makes decisions at the end of life. One must concede that moral attitudes prolong one's existence even if one is dying. Only then it is possible to decide upon active or even passive euthanasia. One could only avoid metaphysical implications in moral attitudes toward someone if one assumes that the human dignity collapses at the end of life. But then there is no demand to respect humans at the end of life, neither dead people nor people who have lost their dignity through suffering. The only way for respecting one's dignity would then be to respect one's life only until to that point in life when dignity is lost. But at this point, suffering at the end of life might not be an ethical problem anymore. And we could mistreat the patients as we wish – simply because they have no dignity anymore. This makes human life purely a technical entity, and there would be no need to respect the factual moral attitudes. The practical consequences of such a position would be horrendous.

Conversely, as long as moral attitudes refer to a moral reality, end of life decisions presuppose a transcendent element of the dignified personality of the patient.

10.4 How metaphysical Dignity works in Ethics

It is obvious that it could violate human dignity if someone is treated harmfully. Although dignity is eternally protected, one could lose his dignity on earth, in life. That is the reason why I have stressed that the term of human dignity entails two different meanings, which nevertheless refer to each other. Thus we must clarify how the relationship between both meanings of dignity should be described. Both meanings refer to each other because the eternal meaning should give an orientation about how we should treat someone under mundane conditions.

Therefore the term of moral attitude is helpful. To respect someone in his eternal dignity means to treat him so that his mundane conditions correspond to his eternal dignity. The moral attitude accompanies the moral agent in his behaviour to humans. However, the moral agent needs a criterion to determine whether he accords to eternal dignity of the other person under mundane conditions. The eternal dignity must be reflected under mundane conditions. Otherwise one would never know whether the eternal dignity is respected or not.

Simply the feeling of the moral agent to respect someone's dignity is not sufficient for respecting human dignity. Although the attitude is helpful, it does not decide if dignity is respected. The attitude of the moral agent opens the window to discover the human dignity of the other. But the criterion for human dignity must be found in the other.

The French philosopher Emanuel Levinas has pointed out in his book "Totality and Infinity" that human dignity entails the idea of infinity. The idea of the infinite is something we could never get unless it reveals itself. Infinity is not at all imaginable. If we want to think about infinity, we could never come to an end because it is infinite. Even it is not sufficient to think the infinite simply by negating the finite. For negating the finite we must already be able to transcend the finite. But this is only possible by experiencing infinity. According to Levinas, I could only think the infinite by transcending my thinking. As long as I am alone with me, with my thinking, with the Cartesian "Cogito", I could never transcend my finitude. I can only transcend my finite thinking and the finite contents of my thinking if I have experienced something out of my self.

Thus Levinas concludes: The infinity is revealed by the Other, the other human. Not other things of the outer world are witnesses to infinity because they could become simple contents of my thinking. But the Other is not at all imaginable from myself. The thought of the Other is excluded in my subjective thinking. Otherness is not subjectively possible, because subjectivity is only self-reference. Consequently, if we think of Otherness, it must be revealed. And since Otherness is revealed, subjective thinking transcends to infinity. Also after revelation of the Other, infinity is unthinkable. But the reference to infinity is possible, to something which is never thinkable. Without the revelation of the Other it is entirely impossible to refer to something unthinkable. Everything which is in my thoughts is also thinkable. Thus I could never have the idea of something which transcends my capacity for thought. Only the revelation of the Other supports the reference to the infinity, although it remains unthinkable.

In his book, Levinas gives reason for the connection between infinity and human dignity. The Other represents the idea of infinity for me. And my impression of infinity corresponds to an attitude that the Other is overwhelming me. His claims are infinite. I could never fulfill them.

I would not like to describe Levinas' ethics on the whole. But his main problem is: How could such a start lead to an ethics? If I was never able to fulfill the Other's claim, because I am never able to fulfill the infinite scope of that claim, ethics would collapse. The system of ethics would fall down entirely. Fortunately, Levinas has found a solution how the Other decreases his infinite claims so that ethical behaviour and even justice to a plurality of Others becomes possible. Here I only want to give a heuristic summary: Levinas describes the Other as connected with the infinite, with a transcendent reality out of myself and out of my life. But he brings together the infinite claim of Otherness with an ethics under finite conditions. The bridge between infinity and (finite) ethics is the Other. The Other is the mirror of infinity but also a revelation in my world. He is the incarnation of the Infinite on earth. Therefore Levinas himself calls his ethics "messianic"[57]. It is the face of the Other which has both directions: the direction to Infinity and to vulnerability: The Other is "naked"[58], he is entirely vulnerable. His infinite claim is to be protected by myself because he is naked.

For an ethics of human dignity it is right to focus on the Other. Human dignity is revealed by the Other. It is not revealed by my own capacities: Humans have not dignity because I have a dignity. This conclusion is not logically valid and it will also fail ethically as Levinas pointed out: The old humanism which has begun with me had never prevented racism, pogroms and genocide – but human dignity is not revealed by God's law. The theological or metaphysical connection of human dignity is its incarnation: The human mirrors God's gift of dignity. The human mirrors infinity. Thus our moral attitude toward human dignity is an attitude toward the human himself.

This sounds trivial but it is not, especially not for theologians. Theologians may think that human beings have dignity because God commands us to treat

---

57 E. Levinas: Totality and Infinity; Kluwer 1979, 285.
58 E. Levinas: Totality and Infinity; Ibid., 129.

each other with dignity. By this understanding dignity is defined heteronomously. Not the human himself would deserve dignity but actually God would deserve worship, and only because of worshiping him we would also respect our neighbour. But think of the highest Commandment which is double-founded. Note the passage of the Gospel: "On one occasion an expert in the law stood up to test Jesus. Teacher, he asked, what must I do to inherit eternal life? What is written in the Law? he replied. How do you read it? He answered: 'Love the Lord your God with all your heart and with all your soul and with all your strength and with all your mind'; and, 'Love your neighbour as yourself.' You have answered correctly, Jesus replied" (Luke 10, 26–28). Actually these are two commandments, but Luke describes them as one. Here is a difference to Marc and Matthew. In these both these Gospels, the love of the neighbour is called "the second" important Commandment (Mt. 22, 39, Mk. 12, 33). Here, in the Gospel of Luke, it is not mentioned which Commandment is the most important. Rather Luke describes how to reach eternal life. Luke deals with infinity: How could I reach infinity? And Jesus' answer refers to a relationship between God and the neighbour, as in the framework of Levinas' thinking.

According to that conception, as in the Gospel or as in Levinas' work, human dignity might not be independent from God, but it is independent of God's Commandment. The opposition is true: God's Commandment refers to human dignity and thus it commands love of the neighbour. The human himself mirrors God's infinity; it needs not be mediated by law. The German theologian Traugott Koch has insisted that the human deserves dignity also in face of God. God must respect human dignity. God does not stand above human dignity. In my eyes, this seems to be imprecise. But the core idea in Koch's assumption is that God cannot contradict human dignity because human dignity mirrors God's infinity. So, if God contradicted human dignity, he would contradict himself.

Note another biblical passage, in which the son of man judges the peoples in Mt. 25. The criteria for his justice is not mediated by his law but consists simply in the situation of the Other: The son of man identifies himself with the humans in their vulnerability: with the hungry and thirsty ones, with the strangers and naked ones, with the prisoners. They themselves in their vulnerable situation represent the glory of the son of man.

Why do I highlight the fact that human dignity depends on the human himself and of nothing outside from him? Because otherwise we would draw out heterogenic definitions of human dignity. And thus we could harm the dignity of a human. Paternalistic dignifying of the other's value harms human dignity. I will not assert that paternalistic acts are generally wrong. Rather I would like to distinguish between paternalistic acts and paternalistic dignifying of human dignity. Not the attitude of the moral agents decides what human dignity is, but the human himself. In other words: Human dignity has to do with the autonomy of the human.

Conversely, I will not say that autonomy of the human *defines* what dignity is. Neither the attitude of the moral agent nor the autonomous self-determination of the human defines what dignity is. The reason is twofold, a principal and a practical one: Firstly (the principal argument): if autonomous self-determination defines human dignity then again it is an immanent definition of the pure subject. And the pure subject cannot transcend itself unless Otherness is revealed. Thus autonomy cannot define what human dignity is.

Secondly (the practical argument): if, for instance, a patient defines what her human dignity consists of, then she would demand that the physician to do everything the patient wants to. Now a dilemma would become apparent: Suppose the patient forces the physician to kill her because otherwise he would harm her dignity. As long as she could define autonomously what dignity is, she could define that she would only be respected in dignity if she was killed. Now the physician could answer to her: If she forces him to kill her she would harm his dignity because he simply has defined so by his autonomous definition. This dilemma would arise if autonomy defines human dignity.

My argument is more careful, more hesitant. I have said: Human dignity *has to do* with the autonomy of the human. Without respecting one's autonomy we would fail to respect his human dignity. Autonomy is a necessary condition of dignity but not a sufficient one. Why? As I have already mentioned in the chapter about human freedom, we must assume freedom for every ethical project. Without the presupposition of freedom there is no ethics. Human dignity, which refers to the human himself, refers to the moral capacity of self-determination. It refers to the human becoming a moral agent. Whoever is allowed to decide for himself what would be morally right for himself, has human dignity. This criterion is valid also for children and mentally handicapped persons, even for persons in coma, and – in my opinion – even for dead persons. This seems to be true at least as an ethical presupposition. We ascribe every person's autonomy insofar they themselves feel their own interests, even if they do not articulate what should happen with them.

The attitude of respecting one's dignity means to respect one's interests. This does not mean that these interests should be fulfilled in general, because their autonomy is not a sufficient condition for their dignity. But it means that we should be careful enough to reject their interests only if there are stronger reasons against this. Therefore, in the German discussion about human dignity, I have proposed combining respect for human dignity with trustworthiness. A moral agent behaves in respect to someone's dignity if his action is trustworthy for the latter. Whether a medical treatment is trustworthy does not only depend on the opinion of the patient. It does not only depend on his factual trust. A patient might factually distrust a treatment but nonetheless it could be trustworthy.

Here again, I want to reaffirm that neither the attitude of the moral agent nor the autonomous self-determination of the patient defines what dignity is. Rather it is a correlation between autonomy of the one person and the implications of respect of the other person. This correlation controls both claims so that they remain in balance with one another. Neither the autonomy of one person could overwhelm the other nor paternalistic actions the first.

Therefore, I have proposed to bind trustworthiness to loving behaviour or at least love-similar behaviour: behaviour which could accompany loving actions. We need not love a person in order to respect her dignity. But in respecting one's dignity we will omit every action which would contradict loving actions. Whoever acts in love will not accept actions as morally right which would harm the dignity of a loved person. Therefore loving people have good instincts concerning what kinds of actions must be excluded in order to respect one's dignity.

But love as only the single criterion could threaten one's dignity because it tends to paternalism: Because I love you, I know better than you what is in your best interest. The conclusion is that we need both criteria: We need love and trust which control each other's moral deficit. Therefore my criterion for respect of human dignity is the following: *"Act for someone as if you love him. And understand love in that way that the other could trust your attitude"*. Neither do you need to love the other in order to respect his dignity, nor the does the other need to trust your attitude factually. But your attitude should be controlled by the test whether the other *could* trust you. Conversely, since the other's claim becomes weakened from autonomy to trustworthiness, you have to control your actions as to whether loving persons could do what you do. Thus trust and love control each other. Dignity happens between the overwhelming autonomy of the human and the paternalistic threat against him.

Note that both concepts – love and trust – also transcend human life, as does dignity. We love persons also after their death – otherwise we could never mourn. And at least trustworthiness is something we ascribe also to dead persons. We fulfill someone's last will after his death because we feel the duty to do so. We feel that fulfilling the last will corresponds to his trust in us although he is actually dead. Thus the concepts of love and trust suffice the transcendence of human dignity.

Now let us turn to the question how my criterion for respect works in end of life-decisions? What should a physician do if the patient wants to die? What should the physician do if the patient asks him to kill her? For answering that question I would like to ignore the laws of different countries. I only focus on the purely ethical question which kind of action would be justified then.

On the one hand, there is the strong claim of the patient: She wants to die. She does not only to want to have essential treatment withheld, but she wants the physician to determine her death, to kill her. We have said that the autonomy of the patient is a strong claim for human dignity. It might not be sufficient for respecting her dignity to fulfill her factual interests. But the physician is not permitted simply to ignore her will to die.

Suppose no one else has an interest how this conflict should be solved. There are no relatives, no juridical limitations, no different interests of physicians. This is not a realistic case but the simplest one. And even in this simple case, we could draw out several ethical conclusions. Perhaps the physician does not feel harmed to kill the patient because he will respect her dignity even after her death. To assist her dying would happen within a framework of transcending the respect of human dignity: The physician could love her also if she has died. I assume he will not love her, but he could. And this is sufficient for respecting her dignity. In that case, killing the patient would be accordant to my criterion of respect: Killing the patient would be trustworthy.

Surely you will have strong objections against my conclusion. But please note that I have argued that human dignity only refers to the human and not to any outer law. Surely one could object that killing contradicts Divine law. I concede that, but my focus is only the concept of human dignity. And therefore the autonomy of the human is relevant as is our trustworthy relationship to him. You could object that the will to die contradicts someone's autonomy. Whoever wants to die is urged by any pressure and is no longer free.

I think this depends on the concept of autonomy. The concept of autonomy must be very rudimentary in order to integrate children, mentally handicapped persons and even dead persons. We have to suppose that everyone knows which interests he or she has. Moreover, remember that freedom has to do with accepting what determines the agent. Thus there might no contradiction between autonomy and the pressure which leads to the wish to die. This rudimentary concept of autonomy underlies the concept of dignity. Otherwise paternalism threatens one's dignity. The consequence would be that we only accept one's autonomy as long as he shares our values and not because he is the one who decides what would be right for him.

What I have in mind is not to justify mercy killing, because stronger reasons might ban this practice. Even if human dignity refers to the autonomous human, it might be that voluntary active euthanasia must be restricted for other reasons. But as long as we only focus on the concept of human dignity, voluntary active euthanasia might be justified.

I agree that there are many other cases which will make the problem more complicated. Others could be frightened if the patient will be killed. Especially handicapped people could feel discriminated or under pressure to decide similarly. Insurance companies could withdraw contracts if treatment of patients who seem to have no benefit become too expensive. Physicians could feel harmed in their self-esteem if they become killers. These are strong reasons to connect the concept of dignity with a concept of justice. Euthanasia might happen in accordance to human dignity but it could be unjust. I will only mention that.

But my summary of that section has to do with the ambivalence of human dignity. Human dignity has several ambivalent structures:

1) The tension between the attitude of respect of the moral agent on the one hand and the autonomy of the human on the other shows that dignity depends on several moral influences, even if human dignity might be the highest moral principle.
2) The correlation between love and trust shows that in human dignity, several values are implied, which must remain in balance so that no one could dominate the other.
3) The ambivalence of dignity as an eternal value shows that human dignity could even trump human life. If dignity is of the highest level, human life is not absolutely protected by human dignity. Dignity could even trump human life.

In the chapter above, I have mentioned that the problem of euthanasia confounds our moral feelings. As we have already seen, it puzzles also ourselves when we discuss about that. The ambivalence in the concept of dignity cause this puzzle. Nevertheless, this does not mean that we should reject the claim of human dignity. We have pointed out that - according to a deontological ethics - the attitude makes actions morally significant. If the attitude of the moral agent refers to the human – as I have pointed out – then it refers to human dignity. Consequently, if a deontological perspective is based on a moral attitude, it could hardly resist the ambivalence of human dignity. We have to live with ambivalence – in ethics as well. And this leads me to the next chapter.

11 Ethical and theological kinds of guilt

In many cases we behave justified, no one accuses us, every moral principle is respected. But nonetheless we feel guilty for having performed in the way we have factually done. We have a bad conscience, although everyone agrees that we have respected the moral order entirely. This might especially happen in cases which have to do with life and death.

In my parish, I often experience that the relatives feel guilty after a natural death of a family member because of his death. They do not feel guilty because there was a failure in treatment. They do not feel guilty because they have harmed him technically or morally. But rather they feel guilty because of symbolic actions which contradict the dignity of the dead person. Let me give some examples: "I was going into the supermarket while he was asleep. After coming home, I noticed that he was already dying. If I had been here he still could have been alive." Here the feeling of guilt arises from a pure absence of the relative during the process of dying. The relative does not feel guilty because she has withdrawn an essential treatment. Perhaps she might have treated him if she has been there, but perhaps she could not. Maybe the process of dying could never be stopped. And maybe an intervention would have demanded a medical know-how the relative does not possess.

The point is not that the relative has failed to do something. Often, we are absent while our family members are sleeping. And we have good reasons to go to the supermarket; it is justified to go there. There is nothing immoral about leaving a family member as long as no one could expect that this family member needs essential help. But during my meetings with the relatives of a dead person, one point is very interesting: It does not help anything to reassure that the relative has no guilt. It does not help to alleviate the guilty conscience. The person feels misunderstood when someone refuses to feel her guilt.

In such cases, I often ask that person what would help her better: Would it help when the dead person could speak to her: "I am going to forgive your guilt". And in almost all cases the person agrees. Not refusing feelings of guilt but forgiving provides comfort. And then the meeting turns to a theological subject: The guilt refers to the fact that we love several people; but our love could not protect them from dying. This is no moral issue anymore but a theological one: It does not deal with moral guilt but with existential guilt: People feel guilty simply because they live but their relative has died.

I have already mentioned that perhaps the relative might not have the medical skill in order to help in such serious situations. But the problem is not that she has no medical skills. Rather the guilt refers to the weakness of her love. Her love is too weak to keep someone alive. It is a problem in principle and has nothing to do with concrete skills. As long as humans are not omnipotent and omniscient they must live with practical deficits. This is not a moral problem but an existential one. It refers to the deficit of human love. Human love is too weak to keep someone alive.

This is the reason why only grace helps in that situation. Grace turns from human inability into a transcendent power. When a transcendent power forgives, this shows that the loving person is held by that power. It is the precise turn of the existential problem the loving person has. She feels guilty that her love could not protect the loved person from dying. But forgiving from a transcendent power protects the loving person from guilt. The point is: If the dead person himself forgives, this would show that the transcendent power is stronger than the capability of human love. The loving person could never be guilty because she has not actually lost the partner anymore. It is the contrary: The dead partner rebuilds the relationship to her from a transcendent power by his grace.

And this is the power of Christian faith. God forgives the sin of the loving person whose love is too weak for eternity. But eternity founds a stronger basis than human love, that is transcendent love, Divine love. By holding the dead person in the kingdom of God, the guilt of the loving relative is forgiven.

I assume that our contradictory feelings on the issue of euthanasia have also to do with the difference between moral guilt and existential guilt (theological guilt). We are confused about which kind of guilt we feel. We are not even aware of that difference and which kind of guilt is valid for our thinking. This is a problem we could not solve for ourselves. Therefore we need consolation from outside. This is the preaching of grace. This is the encouragement of pastoral care. But it is not a task of ethics.

It seems to me that the interest in euthanasia is based on that confusion. Sometimes the relatives hope to solve their existential problem by euthanasia. This sounds absurd, but it becomes clear if we bear in mind the typical context of end of life decisions. In such cases the patient's suffering is burdensome. Often only a rudimentary communication is possible. For the relatives this is a painful situation: They feel helpless and even guilty. It is the same kind of existential guilt, because it depends on love and the inability to hold the patient in life and welfare. Often, the relatives feel that the patient might be still alive but his pure existence has nothing to do with the quality of life a loving person would like to offer. In such cases the relatives often confess that "this is not still a life" or "that life harms the dignity of him". It is against the patient's dignity.

The negative backside of intensive care involves the existential feeling of guilt in the end of life. The loving person suffers because she feels unable to save the life of the loved person. She feels disabled to do anything for him – except killing him. This seems to be the only way to show one's love to the patient: to become a master over life and death. Since the loving relative is a master over life and death, she will not lose the patient after his dying. His death comes within the horizon of her love. He will die because of her love. This changes the existential situation entirely. Whereas in the first case above, the relative feels guilty because her love could not keep the family member alive, now the love becomes responsible for his death. His death becomes an expression of her love. She moves her love into a transcendent area in which the dead person and her love remain connected.

Frankly, this is an illusion. Unfortunately the existential guilt will return after death of the patient. In my parish I experience that after death of someone the relatives are very proud at first when they have decided to withdraw essential treatment. Often family members are ordered as proxies for the patient especially in cases in which the patient is not able to articulate his will. In such cases the relatives might have a first euphoria directly after the death: They were able to do something for the patient, something existential which expresses their deep love: to decide for the patient on the whole – until death. After some weeks the euphoria might be disappeared. And I know many proxies in such families, who feel that their responsibility over life and death is too heavy. Even if there is a first euphoria it disappears after several weeks simply because the suggestive supremacy over life and death does not replace the factual mourning. It is still sad to lose a loved person. And because the loss is felt the mastership over life and death breaks down. The relatives are no masters over life and death, they remain unable to hold a loved person alive. They are unable to keep him in community.

Now the existential guilt returns. And it reveals the illusion of the claim to assist someone dying because of love. It might be an ethical decision to assist someone dying. But this ethical decision does not change the existential problem that even love could not keep someone alive. The crucial point is that at the end of life decisions moral feelings become confused. Existential and moral problems mingle. Existential problems appear to be ethical ones which could be solved by ethical decisions to shorten one's life.

Therefore, theology has to insist in the difference between moral guilt and existential, theological guilt. Never solve existential problems ethically! Even if everything is fulfilled according to ethical claims, the existential problem remains. As confirmed by the biblical passage of the Gospel by Luke: "So you also, when you have done everything you were told to do, should say, 'We are unworthy servants; we have only done our duty'" (Lk. 17, 10). I would like to add: But we have not done what only God is able to do: to keep someone alive simply because of his love.

To sum up: We must distinguish between these two kinds of guilt. We must distinguish between two dimensions of feelings: moral ones and existential, theological ones. And we have to show which dimension is conflicted at the end of life. I agree that euthanasia might be justified in some cases only if we refer only to the ethical dimension. But on the other hand, we might be aware that existential problem is finite in our power and in our love. This problem could never be solved by ethical efforts. Consequently, ethics become demonic since one tries to solve existential problems through ethical decisions. Theology can help for ethics to remain what they are: ethics and nothing else. But if ethics only have such a small task, we should not be surprised when ethics do not solve all problems. We should not be surprised when ethics do not solve all problems entirely. Tragic consequences might also remain even if we have fulfilled all of our duties.

Thus it seems to me that euthanasia might be an ethical option, even active euthanasia. One could withdraw an essential treatment and even kill someone in respect to his human dignity. One could be prepared for mercy killing in accordance to moral attitudes. But one should bear in mind not to mix ethical negotiations with existential problems. One should never use ethical means in order to solve existential problems which could only be solved by God's grace.

My conclusion to call mercy killing an ethical option will not persuade everyone. But it seems to me that the reasons for our objections are based on a different dimension of attitudes and feelings which overrides the ethical dimension. Our main objections against euthanasia do not depend on human dignity or autonomy or other ethical principles. Rather our objections against euthanasia depend on religious attitudes. We have religious objections against euthanasia. And these objections have to do with the existential dimension of human life. But let us bear in mind the difference between ethics and religion, between ethics and faith. Mercy killing might be ethically permitted but nonetheless risky for our existential self-understanding. This is why Christians are reluctant to argue in favour of euthanasia although they believe in life after death. However, their religious sensitivity for the existential dimension of human life makes them hesitate to shorten life too quickly.

12 The Metaphysical Concept of Presumed Will

Especially in Western countries the autonomy of the patient is overestimated. As I have mentioned autonomy could never define human dignity unless in correlation with trustworthy actions which might oppose autonomous self-determinations. But in medical ethics at the end of life, the right to create a living will becomes more and more established. In most Western countries, it is also a juridical duty to respect one's advance directive. An advance directive is a document – usually written – which declares the living will of a person for cases which have not yet happened. A person declares how she will be treated for theoretical situations in the future in which she could not articulate her will.

The justification for advance directives is based on the concept of presumed will. Its most important scope is decision-making about medical treatment. The concept of presumed will should replace the actual will which is absent when a patient has dementia or unconscious because of an accident or an illness. The main aim is to strengthen the patient's will against paternalism of physicians.

Another example for respecting one's presumed will is to order a proxy who represents the will of the patient. The proxy cannot decide according to his own values. But he has to decide as the patient would if he was able to articulate his will.

The proxy has the only task to declare which treatment the patient would prefer if he was able to communicate. Therefore the proxy takes notice of statements the patient has declared in the past. But the past statements are not obligatory. The proxy needs not to accept them word by word, because he has the obligation to interpret them as applied to the present situation in question. Thus the proxy could alter the meaning of the words entirely if he is able to highlight the sense or the character of the patient's presumed *actual* will. The right to alter assertions of the past is entirely justified by the hypothesis that the actual will of a person can change, and the degree of change depends on the personal character of the patient.

Thus the proxy is obliged to protect the *personality* of the patient by discovering his actual will deriving from his personal character. Despite this, the concept of presumed will protects the actual will first and the personality afterward. Otherwise, if the personality was given priority, paternalistic acts would appear to be justified. The conclusion is that the concept of presumed will protects the patient against paternalism by establishing the actual will of the patient.

In my opinion, the priority of the actual will is the main problem of this concept. The problem consists in metaphysical implications which might not be desirable but which are transported unwillingly. The following implications must be presupposed:

1) The patient has an actual will which can be discovered, for instance by the proxy or the court.
2) The process of discovering depends on a kind of communication between the proxy and the patient although the patient is not able to communicate in a trivial kind of sense.
3) There is a dimension of communication which is deeper than the kind of natural and daily communication between humans. The proxy must be able to communicate with the patient in this deeper sense. This kind of communication is a metaphysical one.

My logical reconstruction of this practical concept may be surprising, because it does not presuppose only simple implications. Thus, several objections to my interpretations are possible. Firstly, one could refuse the first implication. One could argue that the concept of presumed will does not entail that unconscious patients could have an actual will, but they could have had a will whose scope has referred to the actual situation like in an advance directive. This objection would confront the subject's activity of *having* a will versus the *content* of a will the subject has published sometimes ago.

I will not enter a long debate in order to reject this objection entirely. Here I only want to defend the first implication briefly in order to highlight the most essential point: Why should we protect the content of a will which has had a subject who is not able to have it anymore? If the subject is unconscious, he would not have a will anymore according to this premise. Why should we treat him then as if his advance directive expresses his actual will, which he cannot have?

It is possible to give the Utilitarian answer that this is the most *practicable* idea. But if we accept ideas of *only* practicable value then the concept of presumed will can only *suggest* to protect someone's autonomy, but it also fails to suggest it because no one really thinks that the patient *has* an actual will. Moreover, paternalism could be as practicable as the concept of presumed will. And we would not have any *moral* reason against paternalism anymore if the concept of presumed will does not actually suppose that the patient has an actual will.

Consequently, the only reason for the concept of presumed will would be that there is any actual will of an unconscious patient. And one must suppose that this actual will is coherent with the presumed will. If there is no actual will the concept of presumed will would not only be arbitrary but it would also fail to *refer* to any real part of the situation in question. This concept would not have any reference to all the problems it should solve. If there is no actual will why should we protect *it* - the *patient's* will which actually would not exist? If we maintain the concept of presumed will we have to think that the patient still has an actual will and that we could respect it. But if this procedure to find one's presumed will is useful, it must imply that there could be real communication between the patient and the proxy. Otherwise the proxy would invent a fiction of the patient's will. And such a fiction would not protect the patient's actual will. Thus, since the first implication is accepted, one would necessarily turn to the other implications, too. This means that one necessarily turns to metaphysical implications.

Remember that I have proposed above that autonomy could not define human dignity. Rather, respect in human dignity depends on trustworthiness in the moral agent respective to his actions. Also the concept of presumed will depends on trust. Conversely, my interpretation of human dignity is also based on metaphysical grounds, but it seems to me that the implications of trustworthiness are more plausible than the metaphysical presuppositions in the concept of presumed will. This is a good reason for a full replacement of the concept of presumed will by the concept of trustworthiness. In my opinion trustworthiness fulfills its demands even if concerned people are incapable of communication. As I have illustrated above, trustworthiness works even if the people concerned are dead. Trustworthiness could be valid also if trust is only virtual. Conversely, there is no possibility to respect someone's actual will if he has none or only a virtual one. This is the decisive advantage of trustworthiness as a basic frame of decision-making. It also functions in cases in which the concept of autonomy becomes derivative or incoherent, such as cases of respecting unconscious people.

Nonetheless one objection could still arise. Whoever fears paternalistic actions by the priority of trustworthiness could argue that the "deeper" dimension of communication which is presupposed in the presumed will needs not really be metaphysical. It could simply be interpreted as transcendence in immanence . The communication between patient and proxy is necessarily embedded in a very natural fact, namely the actual perception of the patient's situation by the proxy. Phenomenological philosophy has generally stressed that primordial perceptions are doubt-free: There is no possible doubt that I perceive what I perceive. Thus if a proxy *feels* spontaneously as the patient does, this would not be a metaphysical way of communication but a kind of empathy within natural communication from which daily communication derives. The "deeper" communication between patient and proxy depends on their so-called „In-der-Welt-Sein" (To-be-in-the-World) like the philosopher Martin Heidegger has called that phenomenological framework. The experience of being aware of the will of someone else is a natural experience, and there is no doubt that one feels what one feels. In this state of phenomenological experience there is no room for skepticism against this feeling of communication.

But it is also clear that skepticism plays an important role to control acts against illusory projection. There is no doubt that I feel communication when I feel communication. And no doubt patient and proxy are interconnected in a communicative relationship even if the patient is unconscious. According to communication theories, communication is independent from action. Even if the patient does not act he could communicate. But it does not follow that the patient distinctly declares his actual will through that communication. It does not follow that the contents of empathy correspond to the real situation of the patient adequately. Thus if a proxy discovers the will of an unconscious patient he will never certainly know if this is his *real* will. But according to the concept of presumed will the proxy must know the real will of the patient in order to claim that this will should be respected. So the concept of presumed will has a different basis than phenomenological truth-claims. It must presuppose that the patient distinctly declares his will by the empathy of the proxy. But this presupposition is based on pure *speculation* of a transcendent reality.

Speculation leaves the phenomenological method and the reference of a phenomenological world. Thus it leaves the world in question and refers to a second world whose transcendence cannot be embedded in phenomenological immanence. Such a second world is independent from mundane properties and must be called metaphysical.

To sum up: The concept of presumed will actually tolerates errors in reconstruction of the actual will. It is apparent that actually the presumed will is only a construction, a projection. But such a projection contradicts the logic of autonomy. Rather such a projection is of the kind of paternalism. Let me compare the concept if it was taken as the real will. Assume a couple. One day the wife confesses: She wants to divorce her husband because she does not want to live with him anymore. According to the concept of the presumed will her, husband could object that she had already decided that she would want to live with him forever. By wedding, she had promised to enjoy living with him forever. Apparently the wife could answer that once she wanted to live with him forever. But she had not decided to *want* to live with him every time. Even by the promise in marriagse one could never anticipate one's own will. One could only promise that now the actual will entails living together forever.

Our promises about the future are possibly erroneous. We could not know whether we will desire later what we desire now. We could never anticipate our prospective actual will. Interviews with patients in the terminal phase of their life or after coma show that people change their interests in the actual situation in ways they could not have foreseen[59]. We all know that. It is our daily experience that our desires change easily. But the concept of presumed will ignores that, in order to find an easy criterion for decision-making. Therefore, we must either develop metaphysics of transcendent communication between patient and proxy. Or we concede that respect for someone's will entails the risk for errors. But then we leave the concept of autonomy and enter the ethical space of trustworthiness.

According to the concept of autonomy, it is not justified to accept such errors. It is not justified that we withdraw an essential treatment when we know that we may only think that the patient would desire to withdraw the treatment. One might object that withdrawing the treatment would be in the patient's best interest. But this is only true in a concept of benefit, not of autonomy. And such a concept of benefit may include paternalistic acts. According to a concept of autonomy, the patient must decide for himself what would be best for him. Consequently, one must concede that the concept of presumed will has no justification for how the principle of autonomy works in case.

I conclude that we accept the risk of making erroneous projections about the actual will of an unconscious patient. We have to accept that. And this means that we could only justify this concept by another reason outside the concept of autonomy. And this is the concept of trustworthiness. To examine the actual will of an unconscious patient is merely an issue of trustworthiness but it is not the doubt-free articulation of his real actual will. It is trustworthy to do so even if the articulation of the presumed will is fallacious. Trustworthiness is an implication of dignity. Thus we could only practice the concept of presumed will, be-

---

59 Chr. J. Ryan: Betting your life: an argument against certain advance directives; Journal for Medical Ethics 22/1996, 95–99; R. Gillon: Autonomy, respect for autonomy and weakness of will; Journal for Medical Ethics 19/1993, 195f.; R. Dresser/P. J. Whitehouse: The Incompetent Patient on the Slippery Slope; Hast.-Cent.-Rep. July-Aug. 24/1994, 6–12.

cause we respect in someone's dignity even if his autonomy and his capacities of autonomous decisions are absent.

Dignity has its own metaphysical implications as I have already pointed out. But these are more plausible because they are more practical. They are well embedded in daily life. Trustworthy decisions include representative decisions for humans who could not decide themselves. The concept of autonomy is misled as long is it should be the whole arrangement of ethical decision-making in a pluralistic world with different value-systems. The metaphysical conditions of dignity may not be plausible for everyone, especially in a pluralistic world. But they are better embedded in presuppositions of daily life than the metaphysics of presumed will.

13 At the End of the Lectures

In these lectures I have proposed several theological links to bio-ethical issues. I have shown that bio-ethical problems often have to do with presuppositions about reality, which are not self-evident. Many bio-ethical problems touch our moral feelings, but seem not to have to do with moral states of the living entities concerned. Or they presuppose metaphysical realities which demand us to think theologically. I have shown that in many bio-ethical issues, there is a useful link for theological ethics.

Theological ethics establishes bridges between Christian belief and pluralistic lifestyles. This is an important function of Christian ethics, one which is not realized by any other theological discipline. In bioethics, theology develops a bridge for revising alternative belief-systems, lifestyles and value-systems. Even ontological frameworks become questionable if theological ethics analyzes their presuppositions in bioethical issues. For building bridges, it is important to understand the consequences of alternative belief-systems, to rethink them, to play with them. Therefore, one needs the creativity for alternative, even contradictory belief-systems. Therefore I would like to invite you to remain on that way of discussing new scenarios of life.

Theologisch-Philosophische Beiträge zu Gegenwartsfragen

Herausgegeben von Susanne Dungs, Uwe Gerber,
Lukas Ohly und Andreas Wagner

Band 1   Walter Bechinger / Uwe Gerber / Peter Höhmann (Hrsg.): Stadtkultur leben. 1997.

Band 2   Elisabeth Hartlieb: Natur als Schöpfung. Studien zum Verhältnis von Naturbegriff und Schöpfungsverständis bei Günter Altner, Sigurd M. Daecke, Hermann Dembowski und Christian Link. 1996.

Band 3   Uwe Gerber (Hrsg.): Religiosität in der Postmoderne. 1998.

Band 4   Georg Hofmeister: Ethikrelevantes Natur- und Schöpfungsverständnis. Umweltpolitische Herausforderungen. Naturwissenschaftlich-philosophische Grundlagen. Schöpfungstheologische Perspektiven. Fallbeispiel: Grüne Gentechnik. Mit einem Geleitwort von Günter Altner. 2000.

Band 5   Stephan Degen-Ballmer: Gott – Mensch – Welt. Eine Untersuchung über mögliche holistische Denkmodelle in der Prozesstheologie und der ostkirchlich-orthodoxen Theologie als Beitrag für ein ethikrelevantes Natur- und Schöpfungsverständnis. Mit einem Geleitwort von Günter Altner. 2001.

Band 6   Katrin Platzer: *symbolica venatio* und *scientia aenigmatica*. Eine Strukturanalyse der Symbolsprache bei Nikolaus von Kues. 2001.

Band 7   Uwe Gerber / Peter Höhmann / Reiner Jungnitsch: Religion und Religionsunterricht. Eine Untersuchung zur Religiosität Jugendlicher an berufsbildenden Schulen. 2002.

Band 8   Walter Bechinger / Susanne Dungs / Uwe Gerber (Hrsg.): Umstrittenes Gewissen. 2002.

Band 9   Susanne Dungs / Uwe Gerber (Hrsg.): Der Mensch im virtuellen Zeitalter. Wissensschöpfer oder Informationsnull. 2004.

Band 10 Uwe Gerber / Hubert Meisinger (Hrsg.): Das Gen als Maß aller Menschen? Menschenbilder im Zeitalter der Gene. 2004.

Band 11 Hubert Meisinger / Jan C. Schmidt (Hrsg.): Physik, Kosmologie und Spiritualität. Dimensionen des Dialogs zwischen Naturwissenschaft und Religion. 2006.

Band 12 Lukas Ohly: Problems of Bioethics. 2012

www.peterlang.de

www.ingramcontent.com/pod-product-compliance
Ingram Content Group UK Ltd.
Pitfield, Milton Keynes, MK11 3LW, UK
UKHW021823140426
5217IPUK00004B/59